What Every Engineering College Student Should Know

Be Competitive

What Every Engineering College Student Should Know

Be Competitive

M. Adithan

ATLANTIC
PUBLISHERS & DISTRIBUTORS (P) LTD

Published by
ATLANTIC
PUBLISHERS & DISTRIBUTORS (P) LTD

7/22, Ansari Road, Darya Ganj, New Delhi-110002
Phones : +91-11-40775252, 23273880, 23275880, 23280451
Fax : +91-11-23285873
Web : www.atlanticbooks.com
E-mail : orders@atlanticbooks.com

Branch Office
5, Nallathambi Street, Wallajah Road, Chennai-600002
Phones : +91-44-64611085, 32413319
E-mail : chennai@atlanticbooks.com

Printed in India at Nice Printing Press, A-33/3A, Site-IV, Industrial Area, Sahibabad, Ghaziabad, U.P.

What do you see?

It depends on what do you want to see.

Engineering

More than a discipline or a body of knowledge, engineering is a verb, an action verb, denoting a way of approaching a problem.

—Scott Whitmire

Preface

Engineering colleges and technical institutions in India are an important segment in the Indian higher education system. They have made tremendous impact on and contribution to the Indian economy. India is making giant strides through its expanding and invigorating educational system to assume a significant role in the international affairs.

To promote continuing engineering education, research and innovation in their professional life, engineers and technologists need to be trained while they are pursuing their degree (B.E. or B.Tech.) programme in engineering colleges or technical institutions. The book will give the desired inputs and exposure to them in the early stages of their engineering education and training.

Engineering college students require training in a number of skills to complement their knowledge gained in the chosen professional field of engineering. Four years of study in an engineering college should be fully engaging, fruitful, enjoyable, memorable, professionally enriching and rewarding. It should prepare the students with necessary competencies, not only for the immediate job or career opportunities but also to equip them to meet the economic, social and technological challenges of the future.

The book will impart to the students various types of such skills. It is designed to provide them the knowledge of how to develop communication skills; how to balance study, recreation and relaxation; how to manage time; how to develop positive thoughts for rewarding actions; how to achieve academic integrity and ethics; how to prepare adequately for tests; what are the employment opportunities and how to go for job search; how to learn problem-solving techniques; and what is leadership and how it helps in one's career. It also explains important academic terms. Besides, the book contains five Appendices on pertinent areas.

It is sincerely hoped that the engineering college students will be highly benefited by this book.

M. Adithan

Acknowledgements

I would like to thank Dr. Raymond B. Landis, Dean Emeritus, College of Engineering and Technology, California State University, Los Angeles, U.S.A. for making available the slide presentation, "Studying Engineering: A Road Map to a Rewarding Career, 3rd ed., April 2007", for wider dissemination amongst engineering college students and faculty. Some portions of this text/book have been based on and adapted from this work/reference.

I am also thankful to Prof. Dr. Robert C. Creese, West Virginia University, U.S.A. for the inspiration and guidance he provided in the development of this book.

M. Adithan

Contents

1

What is Engineering?

What is Engineering? How would you Respond?

Engineering is the profession in which the knowledge of mathematical and natural sciences (physics and chemistry) gained by study, experience and practice, is applied with judgment to develop ways to utilise, economically, the materials and the methods for the benefit of mankind.

The Engineer and the Scientist

Both the engineer and the scientist are thoroughly educated in the mathematical, physical and natural sciences, but the scientist primarily uses this knowledge to acquire new knowledge, whereas the engineer applies the knowledge to design and develop usable devices, machines and equipments, structures and processes. In other words, the scientist seeks to know, the engineer aims to do.

Greatest Engineering Achievements of 20th Century

- High performance materials
- Nuclear technology
- Laser and fiber optics
- Petroleum and gas technology
- Health technology
- Household appliances
- Imaging technology
- Internet and e-mail

- Space exploration
- Interstate highways
- Refrigeration and air-conditioning
- Telephone, mobile/cellular phones
- Computers and super computers
- Agricultural mechanization
- Radio and television
- Electronics
- Safe and affordable water
- Supersonic aircrafts
- Power plants and electrification.

Top 11 Technologies of the Decade

No. 1 — Smartphones
No. 2 — Social Networking
No. 3 — Voice Over IP
No. 4 — LED Lighting
No. 5 — Multi-core CPUs
No. 6 — Cloud Computing
No. 7 — Drone Aircraft
No. 8 — Planetary Rovers
No. 9 — Flexible AC Transmission (FACT)
No. 10 — Digital Photography
No. 11 — Class-D Audio.

Rewards and Opportunities of an Engineering Career

- Job satisfaction
- Variety of job opportunities and career options
- Challenging work
- Intellectual development
- Social impact
- Financial security and good salary

- Prestige
- Professional environment
- Understanding how things work and product development
- Creative thinking
- Critical thinking.

Major Engineering and Technology Disciplines

- Civil Engineering
- Mechanical Engineering
- Electrical Engineering
- Electronics Engineering
- Chemical Engineering
- Industrial Engineering
- Aerospace Engineering/Aeronautical Engineering
- Biotechnology
- Information Technology
- Agricultural Engineering
- Automobile Engineering/Automotive Technology
- Communication Engineering/Telecommunication Engineering
- Petroleum Engineering
- Systems Engineering
- Nuclear Engineering
- Space Science and Technology
- Mining Engineering
- Armament Technology.

Employment Opportunities

- Professional, scientific, and technical services
- Design, research, and development
- Information services
- Construction

- Retailing, sales, and marketing
- Administrative, and support services
- Management of companies
- Small business development
- Testing, quality, and certification
- Computing and software development
- Mining
- Repair, maintenance, and re-conditioning services
- Welding and fabrication
- Teaching
- Consulting.

Engineering Job Functions

- Analysis, Simulation, and Modeling
- Design
- Prototype Development
- Testing
- Construction and System Integration
- Sales, Marketing, and Advertising
- Research and Development
- Business and Management
- Consulting/Consultancy Services
- Teaching/Technical Training
- Entrepreneurship.

Functions of Engineering Professionals

Within a given engineering field of specialisation there are many functional areas in which engineers are engaged. These are:

- **Research and Development:** Seeking new knowledge for better understanding and development of new equipments, products, processes, technologies and methods.

- **Design:** Converting the concept into detailed plan and drawing of specifications of equipments, machines, and materials.
- **Production/Manufacturing:** Involving machines, equipments, and processes by which products/ components are made from raw materials.
- **Construction:** The process of converting design into structures, buildings, bridges, highways, machines building, power plants, transmission lines, and communication facilities, erection using appropriate equipments, materials, and processes.
- **Operations:** The application of engineering and management principles in carrying out manufacturing, inspection, procuring of engineering items, assembly, performance testing, directing and controlling of operating personnels, servicing, and maintenance.
- **Sales and Marketing:** Understanding customer' needs, supplying appropriate equipment/product and spares, providing after sales service and maintenance.
- **Management:** Involving policy planning, corporate affairs, finance, organizational structure, public relations, recruitment, selection and training of engineers and managers, and coordination of R&D, production, sales and marketing and ensuring profitability.

Criteria for Engineering Education

(a) An ability to apply knowledge of mathematics, science and engineering.

(b) An ability to design and conduct experiments as well as to analyse and interpret data.

(c) An ability to design a system, component or process to meet the desired needs.

(d) An ability to function on multi-disciplinary teams.

(e) An ability to identify, formulate and solve engineering problems.

(f) An understanding of professional and ethical responsibility.

(g) An ability to communicate effectively.
(h) A broad education necessary to understand the impact of engineering solutions in a global and social context.
(i) A recognition of the need for and an ability to engage in life-long learning.
(j) A knowledge of contemporary issues.
(k) An ability to use techniques, skills, and modern engineering tools necessary for engineering practice.

The Engineering Design Process

The steps involved are:

- Identify the needs or requirements of the customer.
- State the problem, definition, and the specifications of the equipment, machine or system.
- Collect the relevant data and information.
- Develop various alternative designs.
- Evaluate various alternative solutions.
- Select the optimal design or the design that meets your requirements.
- Implement the optimal design.

Eight Dimensions of Product Quality

1. Performance
   - The primary operating characteristics of the product or service.
2. Features
   - The characteristics that supplement the basic functioning of the product or service.
3. Reliability
   - Probability of the product or service failing within a specified period of time.
4. Conformance
   - The degree to which a product or service meets acknowledged standards.

5. Durability
   - A measure of product life (both technical and economic).
6. Serviceability
   - The speed, courtesy, competence, and ease of repair or recovery.
7. Aesthetics
   - How a product or service looks, feels, sounds, tastes, or smells.
8. Perceived Quality
   - Various tangible and intangible aspects of the product from which quality is inferred.

Six Sigma

Six Sigma is a quality control tool. Six Sigma is the most widely used strategy for statistical quality assurance (SQA) in industries today. This concept was originally popularised by Motorola in the 1980's. The Six Sigma strategy is a vigorous and disciplined methodology that uses data and statistical analysis to measure and improve an industry's operational performance by identifying and eliminating defects in manufacturing and service-related processes.

Six Sigma is derived from six standard deviations, i.e., $\pm 3\sigma$ (equivalent to 3.4 defects per million occurrences)—indicates an extremely high quality standard. Higher the sigma level, the lower is the defect rate and lower the defect rate, the higher is the quality.

The Six Sigma methodology consists of the following five core steps:

1. **Define** customer requirements deliverable through well defined methods of customer communication.
2. **Measure** the existing processes and their output to determine current quality performance by collecting defects metrics.
3. **Analyse** defect metrics (defect analysis) and determine the vital few causes.

4. **Improve** the processes by eliminating the root causes of defects.
5. **Control** the processes to ensure that future work doesn't reintroduce the causes of defects.

These steps are referred to DMAIC (Define, Measure, Analyse, Improve, and Control) methodology.

Six Sigma strategy can be adopted in many areas of business/ industry:

1. Product Design and Design Methodology
2. Continuous Process Improvement
3. Continuous Product Improvement
4. Enhanced Customer Satisfaction
5. Supply Chain Improvement/Supply Chain Management.

Six Sigma paves way for the enhancement of human elements such as Strong Leadership, Training and Development, Customer Focus, Project Teams and Culture Change.

Problem Solving

The steps involved in problem solving are:

(i) Identify the problem.
(ii) State the problem clearly in unambiguous terms.
(iii) List constraints, assumptions, and facts.
(iv) Generate possible solutions.
(v) Determine the most likely solution.
(vi) Analyse the solution.
(vii) Synthesize the solution.
(viii) Evaluate the solution.
(ix) Prepare a report.
(x) Implement the plan.
(xi) Check for results.

The above sequence provides a clear process for thinking through the complexities of real problems.

2

What is Success?

You Can Achieve Success!

- Academically poor students and average students have succeeded.
- Highly qualified students have failed.
- Then what makes the difference?

Success is the Achievement of Something

- Desired,
- Planned, or
- Attempted.

Tips for Success (Reasons for Success)

Research on 1500 prominent men and women, who were known not because of their wealth or social position but because of their achievements, quoted the tips for success:

- Commonsense,
- Knowing your field,
- Self-reliance,
- General intelligence, and
- Ability to get things done.

In addition:

- Importance of marketing yourself,
- Developing leadership qualities,

- Maintaining dead lines,
- Working as a team, and
- Learning through criticism.

Goal Setting

How can you expect to get somewhere if you don't know where you want to go?

- Goal setting is important.
- Goals give you something to measure yourself.
- Goals give you direction for your life.
- Hence, you should know your goals and write down your goals as clearly as possible.

How Can You Strengthen Your Commitment?

- By clarifying your goals, i.e., explain your goals in clear terms.
- Learn as much as you can about engineering and technology.
- Prepare a road map which you can follow.
- Work hard to achieve your goal.

Success in Engineering Study Depends on

- Your Efforts—"Work Hard",
- Your Approach—"Work Smart", and
- Your Attitude—"Think Positively Always".

Ability Vs Effort

- Do you believe that people succeed because of their ability that some people "have it" while others don't?

Or

- Do you believe that people succeed because of their effort?

Which do you think is more important—Ability or Effort? Why?

Fixed Mindset Vs Growth Mindset

People generally have one of these "Mindsets".

Fixed Mindset: You believe that your talents and abilities are already set in you—either you have them or you don't have them.

Or

Growth Mindset: Talents can be developed and superior talents and greater abilities can be built over a period of time as one grows.

Both of these mindsets are important and are required for success in engineering study.

Why Students Go to College/University?

(Question posed: Why did you decide to go to College/ University?)

	Reasons	% Responses
1.	To improve my job opportunities.	(79%)
2.	Personal Interest—I love learning new things.	(60%)
3.	It seemed logical to go on from school to college/University.	(45%)
4.	To help develop/change my current job/career.	(40%)
5.	To get away from home and live my own life.	(23%)
6.	I am at the college/university because my family/friends expected me to go.	(18%)
7.	Other reasons.	(5%)

Note: Reasons cited are not mutually exclusive.

Virtues of a Good Student

1. Should possess good communication skills.
2. Should be on time, punctual, and attend classes regularly.
3. Should be interested in pursuing a career in arts, law, science, commerce, engineering, technology, management or business, etc.

4. Should understand the feelings of his/her classmates and other fellow students.
5. Should be a good team player.
6. Should be creative and innovative.
7. Should be a life-long learner.
8. Should respect his/her elders and parents.
9. Should be open to criticism.
10. Should respect his/her teacher for his/her knowledge, experience, and feelings.
11. Should be ethical, moral, and a socially responsible citizen.

How Should You View Your Education in the Engineering College?

- What knowledge, skills, and attitudes will you acquire when you graduate?
- You should know what employers look for in new engineering graduates.
- What can you do to ensure that you get a quality education during your engineering study?

When You Graduate from Engineering College You Should Acquire

1. An ability to apply knowledge of mathematics, science, and engineering.
2. An ability to design and conduct experiments as well as to analyse, and interpret data.
3. An ability to design a system, component, or process to meet desired needs, i.e., functional requirements.
4. An ability to function in multi-disciplinary teams.
5. An ability to identify, formulate, and solve engineering problems.
6. An ability to communicate effectively.

7. An ability to use the techniques, skills, and modern engineering tools (e.g., relevant Softwares) necessary for engineering practice.
8. A broad education necessary to understand the impact of engineering solutions in a global and societal context.
9. An ability to "learn how to learn", i.e., engagc yourself in life-long learning.
10. A knowledge of contemporary issues, i.e., issues concerning present generation.
11. An understanding of professional and ethical responsibility.

Student Aims

Students should aim to:

- engage with the academic, social, and cultural life of the University.
- utilise the opportunities available in the campus to nurture his/her artistic abilities and engage in creative pursuits and take part in extra-curricular activities.
- contribute to the reputation and development of the University by communicating where improvement can be made, by participating in its decision-making processes wherever possible.
- participate actively and positively in the teaching-learning environment.
- beware of, understand and comply with the University's Rules and Regulations relating to students.
- comply with workload expectations, and notify appropriate faculty/mentor if difficulties are experienced.
- respect the Rules and Regulations about malpractices, plagiarism, fabrication, or falsification of data, as well as the proper use of copyright material.

Together, the University and the student body will strive to ensure that every student has been given the opportunity to enjoy a fulfilling and rewarding learning experience which will promote their academic and personal growth.

Three Basic Skills

According to Tony Wagner, Harvard based education expert and author of "The Global Achievement Gap", three basic skills that students should acquire if they want to thrive in a knowledge economy are:

1. The ability to do critical thinking and problem-solving.
2. The ability to communicate effectively.
3. The ability to collaborate.

Ten Life Skills Which an Engineering Student Should Acquire and Develop

- Empathy (putting oneself in another person's state of mind to understand his/her problems)
- Problem-solving
- Interpersonal relationship
- Creative thinking
- Critical thinking
- Coping with stress
- Managing feeling of emotion
- Self-awareness
- Decision-making
- Effective communication.

What Your Employers Want When They Recruit You for Employment?

1. In addition to your qualifications, they look for maturity, initiative, enthusiasm, poise (balanced approach), good appearance, integrity, flexibility, and ability to work with people (Team work).
2. Good grades in all subjects in your branch of engineering.
3. Whether you underwent any specialised course(s) relating to a particular field of work?
4. Ability to communicate effectively, both orally (Spoken English) and in writing (Written English).

5. Kind of industrial training or practical training you underwent while studying in engineering college.
6. Experience in campus activities, especially participation and involvement in co-curricular activities (Seminars, Symposia, Workshops, Conferences, and Professional Societies) and extra-curricular activities (Sports, Games, Cultural events, Students' Club activities).

Your Involvement as a Student in Campus Life

1. Your participation in various students organizations.
2. Your interaction with faculty members.
3. Your interaction with fellow students.
4. How have you spent your time and energy in the campus apart from studying?

How was Your Life Spent in Campus?

1. Stayed in hostel, or a day-scholar, arrangements you have made for living while studying.
2. Influence of family, parents, and relatives.
3. Influence of class teachers/professors.
4. Influence of others, mentor, student adviser, project guide, proctor, warden.

3

What is Learning?

Learning is the Process of Acquiring

- New knowledge and intellectual skills (Cognitive learning).
- New physical skill or manual skills (Psychomotor learning).
- Appropriate emotional responses, attitudes, and values (Affective learning).

How do we Learn?

- By receiving new knowledge, and
- By processing new knowledge, and retaining it in our mind.

Receiving New Knowledge

- How do you like to receive the new information—through sensing or by intuition (i.e., without any reasoning)?
- Through which sensory channel do you perceive external information most effectively—by visual learning or by verbal learning?

Processing New Knowledge

- The way you prefer to process new information:
  - By active learning (by engaging in an activity), or

- By reflective learning (learning by a thought process by engaging your mind).

- The way you understand:
  - You want to learn things in a logical sequence, or
  - You want to learn by forming a mental picture of the entire information.
- Study the information given earlier and think of them so that you know about yourself, i.e., how do you like to identify yourself?

  Whether you learn

  (a) By sensing or by intuition?

  (b) By visual means or by verbal means?

  (c) By active learning or by reflective learning?

  (d) By a sequential process or by thinking of a global picture?

  Answers to the above questions will tell you about your preferred learning style(s).

How Students Learn?

Different students learn in different ways.

Some students

- think symbolically in words and numbers, while others are thinking in the form of pictures and images.
- are analytical, preferring to focus on details, while others focus on the "whole picture".
- are intra-personal, preferring to work alone while others are interpersonal, preferring to collaborate or work in a group.
- are reflective, preferring to think about new information while others are active, preferring to do something (writing or solving a problem) and learn.

Improve Your Learning Process

- Observe your learning style.
- Think about what you have observed and about your learning style.

- Think about how you can improve your learning process continuously.

The Determinants of Efficient Learning

Learning efficiency depends on three factors:

(1) The material to be learned,

(2) The psychological state of the learner, and

(3) The learning strategies.

The more meaningful the material to be learned, the more easily it is learnt (understood). Material that can be related to what is already known is more likely to be remembered.

Any new material can be learned in stages. More practice leads to better performance. Writing the material studied or learnt is better than merely reading (or hearing) it.

The novelty of the material (the extent to which the material is new and interesting) can also influence the learning process. The novelty in the learning process (i.e., new and alternative learning approaches) also increases the attentiveness and learning efficiency.

Students should use new and innovative approaches in performing their home assignments and in solving of problems.

Any anxiety and fear of failure leads to poor performance. Hence, anxiety and fear of failure should be removed from student's mind for better learning and improved performance.

Benefits of Knowing Your Personality Type and Learning Styles

You should know whether you are:

- An extrovert or introvert.
- Sensing a situation or knowing it by intuition (without reasoning).
  – A thinking person or a person with more feelings.
  – Judging or perceiving (becoming aware of).
- This will help you in selecting the most appropriate engineering job that is suitable for you.

- Help you in creating your own learning experiences to meet your needs.
- Assist you in appreciating your own uniqueness and strength.
- Assist you in appreciating the uniqueness and strength of others in the organization.

Characteristics of Good Learners

- Understand your learning process and try to improve your learning process continuously.
- Good learners are active (not passive) in their approach to learning process.
- Good learners are motivated (e.g., they enjoy learning, they have set short-term and long-term goals).
- Good learners are disciplined (e.g., they follow good study habits and use them consistently).
- Good learners are more aware of themselves (as learners they know their strengths and weaknesses).
- Good learners learn more by utilising the opportunities available to them (interacting with their colleagues and professors and by reading good text-books).
- Good learners set specific learning goals for themselves and work towards achieving the same.
- Expert learners have many learning strategies/styles from which to choose.
- Know not only what to learn, but also how to learn (i.e., "learn to learn").
- Systematically plan their approach to learning.
- Monitor their learning process while it's happening.
- Are more adaptive, because they do self-monitoring while learning.
- Think and act upon their own learning style.
- Evaluate the effectiveness of various learning approaches and strategies and choose the best approach.
- Desire to achieve their goals by developing their competencies.

4

The Beginning of the Semester

Following are the activities that you should do at the beginning of each semester.

- Collect your course syllabus/know the titles of the subjects you will be studying in this semester.
- Note down the time table of your class/branch.
- Acquire the relevant/prescribed textbooks and other materials, design data book, if any, charts, problem sets or data sheets you may use in your tutorials, etc.

Preparing Yourself for Attending Lectures

- Review you class notes, read the chapters in the text, solve the problems, answer the questions listed at the end of the chapter(s).
- Sometimes lectures may be like a reinforcement to the subject matter you have already studied/learnt.
- Everyday you should study a small portion of the subject matter and acquire the relevant knowledge (do not postpone your studies to a latter day).

During the Class Periods

- As far as possible sit at the front; you will hear the teacher better and see the board clearly.
- Concentrate on what the teacher is teaching/doing.
- Develop good listening skills.
- Take notes systematically.

- If you have any doubt ask questions in the class and clarify your doubts.

Characteristics of a Good Listener

What you should do:

- Consider that all topics are equally important.
- Listen so that you acquire new knowledge/information.
- Judge the value of the content rather than the style of the delivery of the teacher.
- Listen and identify the central theme. This will help you to build anchor points to follow the entire lecture.
- Be alert and listen carefully what the teacher says.
- Note down the important points discussed in the class.
- Avoid distractions and concentrate on what the teacher is saying.
- Consider difficult portion discussed as an exercise for the mind.
- Listen to the teacher with an open mind.
- Think more deeply about what the teacher is saying, try to summarize the topic that has been covered.

How Should You Take Notes?

- Good notes help you to know what is important.
- As far as possible, use spiral notebook in taking down notes in the class instead of a rigid/bound register note book or stapled note book.
- Listen to the teacher carefully and note down the important points or ideas discussed.
- Draw the diagram, note down the equation or formula the teacher writes on the board.

Asking Questions in the Class

Teachers may ask different types of questions during the class.

- Memory level questions, questions involving Lower Order Thinking Skills (LOTS).

- Convergent thinking questions (All students think and give more or less the same type of answers).
- Divergent thinking questions (Students give a variety of possible alternative answers/solutions which may be considered true or correct).
- Evaluative thinking questions, questions involving Higher Order Thinking Skills (HOTS), like critical analysis, critical thinking, creative thinking.

Professors and Teachers are important resources and they are costly resources which the Institute/College has acquired.

What Professors Do?

Teacher's Roles:

- To provide instructions to the students.
- Academic advising, providing career guidance/ counselling and personal advice to the students.
- To monitor students' progress; and to involve them actively in the teaching-learning process.
- To assess students' work/assignments/answer scripts and award a Grade consistent with the students' academic performance.
- To involve the students in research and development (R&D) activities to the extent possible depending on their abilities and interest.
- To serve as a reference/referee when students apply for a job or higher studies.
- To recommend them for scholarships or for academic awards or for a job or for their higher studies.
- They are engaged in teaching, tutoring and in academic advising.
- They continue to study, learn, and acquire new knowledge.
- They are engaged in research and development (R&D) activities.

- They provide guidance and counselling service/serve as proctors to students.
- They guide their students in various professional societies activities, clubs, co-curricular and extra-curricular activities.
- Infact, teachers perform multiple roles.

Characteristics of Your Professors

- Professors believe that their areas of subject matter specialty, their experience, and their expertise are important to them.
- They are interested in advancing the knowledge in their areas of specialization by carrying out research and development (R&D) and publishing papers in international and national journals and presenting papers in conferences/symposia.
- They have chosen an academic career over professional practice or working in a company/organisation industry.
- They believe they are outstanding teachers and have a passion for teaching young students.
- They are very knowledgeable, and love to convey what they know to others and derive satisfaction in the process.
- They love to interact with their students.

5

Attending Lecture Classes

- Each teaching method involves a faculty member who is knowledgeable about a subject (an "expert") communicating what he or she knows to a less knowledgeable person (viz. the student).
- Generally in lectures, most of the communication is one way, i.e., from the teacher to the student.

Observe How Information is Presented in the Class

1. **What type of information is emphasised?**

   Concrete—Facts, data, observable phenomenon, or

   Abstract—Principles, laws, concepts, theories, mathematical models.

2. **What mode of presentation is stressed?**

   Visual—Pictures, diagrams, sketches, video films, demonstrations, or

   Verbal—Spoken words, words written on the board.

3. **How is the presentation organised?**

   Deductive—Start with fundamentals; proceed to applications, state the reasons, proceeding from general to particular, or

   Inductive—Start with applications; proceed to fundamental (General inference from particular instance or example).

4. **What mode of student participation is facilitated?**

   Active—Student is involved (talking, doing, reflecting on what he learns, solving problems, drawing a diagram, or a sketch, calculating), or

   Passive—Student as a spectator (watches, observes, simply listens).

5. **What type of perspective is provided on the information presented?**

   Sequential—Step by step progression of ideas, or

   Global—Bringing out the full picture, giving the contents and its relevance to the idea discussed.

The Learning Process

Principles of Learning: Carl Rogers (1902-1987) carried out research on learning process, learners' enthusiasms, and has developed a number of principles about how people learn. The principles of learning are:

1. All human beings have a potential for learning.
2. Significant learning takes place when the subject matter is perceived by the learner as having a relevance for his/her own purpose, when the individual has a goal he/she wishes to achieve and sees the material presented to him/her as relevant to the goal, and learning takes place with great ease and quickly.
3. Learning results in self-organization of the matter taught.
4. Learning is better when external threats are kept to a minimum.
5. Significant learning occurs by actually doing and by practice.
6. Learning is facilitated when the learner participates responsibly and actively in the learning process.
7. Independence, creativity and self-reliance of the learner are enhanced when self-criticism and self-evaluation are carried out by learner himself/herself and evaluation by others is of less importance.

8. The most useful learning in the modern world is the learning of the process of learning (i.e., learn-to-learn) (Open to experience and to incorporate into oneself the process of change).

Learning is a Reinforcement Process

When (Situation)	What to do
Before the class	Prepare for the next lecture by reviewing previous class notes, reading the text material, attempting a few problems and solving some questions which will give you confidence.
During the class	Attend the lecture, concentrate on what the teacher does, take detailed notes, ask questions and get your doubts clarified.
After the class is over, but before the next class	Review and read the notes, read the relevant text; complete the problems, choose a study partner, if required, or study group to clarify your doubts. If you have doubts still, meet your teacher and get your doubts clarified.
For preparation for class test	Review your notes; review the text, work out the problems, study the important definitions, workout the extra problems and answer the questions in the text.
For preparation for final examination	Review notes, read the text, work out the problem, read the assignments given and get your doubts clarified; prepare memory cards (Cue cards) which contain important definitions, formulae, and procedures given by the teacher.

You Should Know How Your Professors are Teaching?

- Your Professors use a variety of teaching styles. Probably lecture method is the most common way by which they teach you.
- There are many other methods by which your Professors will also be teaching you.

These are:

1. Quiz
2. Role-play
3. Small group discussion
4. Videos
5. Games
6. Use of models and demonstration kits
7. Cross-word puzzles
8. Projects
9. Questions and answers sessions.

- Whatever method is adopted you should take active part.

Cone of Experience

Edgar Dale (1900-1985; Ohio State University, Columbus, Ohio, U.S.A.), the father of Modern Media in Education, developed the "Cone of Experience" from his experience in teaching and his observations of learners.

As you go down the Cone of Experience, you are involved in Active Learning by doing the things actually, and in this process you remember/retain 90% of what you say and do.

Hence you should be more and more involved in activities such as discussion, group work, group discussion, in giving a speech, doing a role play, fabrication of a device, circuit, system, and in giving a seminar/presentation.

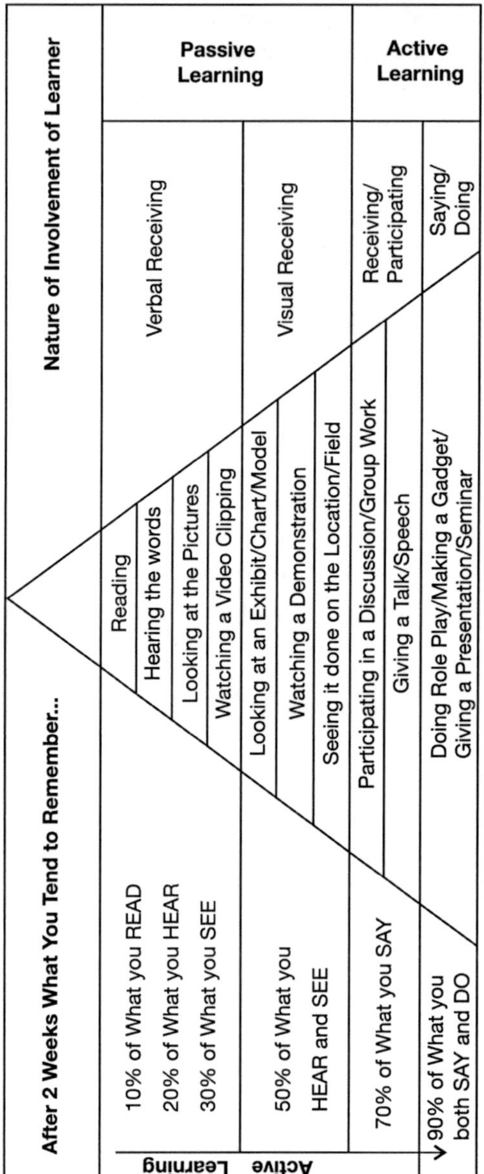
Edgar Dale's Cone of Experience
After 2 Weeks What You Tend to Remember...
Nature of Involvement of Learner
10% of What you READ
20% of What you HEAR
30% of What you SEE
50% of What you HEAR and SEE
70% of What you SAY
90% of What you both SAY and DO
Reading
Hearing the words
Looking at the Pictures
Watching a Video Clipping
Looking at an Exhibit/Chart/Model
Watching a Demonstration
Seeing it done on the Location/Field
Participating in a Discussion/Group Work
Giving a Talk/Speech
Doing Role Play/Making a Gadget/
Giving a Presentation/Seminar
Verbal Receiving
Visual Receiving
Receiving/
Participating
Saying/
Doing
Passive Learning
Active Learning
Active Learning

Levels of Intellectual Skills as per Bloom's Taxonomy

- Knowledge, knowing, and remembering
- Comprehension and understanding
- Applying (Application of concepts, principles, and rules)
- Analysing (critical analysis, critical thinking)
- Synthesis (creating something new, creative thinking)
- Evaluation and judgment

Blooms Taxonomy

Benjamin Bloom created this taxonomy for categorizing different levels of competency to be developed in learning a subject. The taxonomy also provides a useful structure for categorizing test questions.

Competence	Skills Demonstrated	Question Cues		
Lower Order Thinking Skills (LOTS)				
Knowledge	• Observation and recall of information (from memory) • Knowledge of dates, events, places • Knowledge of major ideas • Knowing of subject matter	List Define Tell Describe Identify	Show, Label Collect Where? Tabulate	Quote, Name Who? When?
Comprehension	• Understanding information • Grasp meaning • Translate knowledge into new context • Interpret facts, compare, contrast • Order, group, infer causes • Predict consequences	Summarize Describe Interpret Contrast	Predict Associate Distinguish Estimate	Differentiate Discuss Extend
Higher Order Thinking Skills (HOTS)				
Application	• Use information • Use methods, concepts, theories in new situations • Solve problems using required skills or knowledge	Apply Demonstrate Calculate Complete Illustrate	Show Solve Examine Modify Relate	Change Classify Experiment Discover

(*Contd...*)

Analysis	• Seeing patterns • Organisation of parts • Reorganisation of hidden meanings • Identification of components	Analyse Separate Order Infer	Connect Classify Arrange Divide	Compare Select Explain
Synthesis	• Use old ideas to create new ones • Generalise from given facts • Relate knowledge from several areas/ disciplines • Predict, draw conclusions	Combine Integrate Modify Rearrange Substitute	Plan Create Design Invent What if?	Compose Formulate Prepare Generalise Rewrite
Evaluation (Judgment)	• Compare and discriminate between ideas • Assess value of theories, presentations • Make choices based on reasoned argument • Verify value of evidence • Recognise subjectivity	Assess Decide Rank Grade Test	Recommend Convince Select Judge Measure	Discriminate Support Conclude Compare Summarise

6

Mistakes Students Make

Mistakes Students Make	Strategies for overcoming them (What you should do to avoid these mistakes?)
1. Assume engineering study will be like studying 10+2.	Understand and adjust to the differences between 10+2 and engineering study.
2. They think they will fail. (Losing self-confidence)	Devote adequate time and energy to your studies and develop self-confidence.
3. Spend more time outside the campus.	Engage yourself in the academic environment of the institution.
4. Neglect studying.	Devote time and energy for your studies. Think that education is an asset and is an investment for your future.
5. Delay studying until the test is announced.	Start studying from the beginning of the Semester. Master the topic of each class prior to next class.
6. Trying to study 100% alone by oneself.	Study collaboratively with other students also. Engage in group study, it might help you. Study with your friends and class mates.

(Contd...)

7. Come to the lecture unprepared.	Review class notes, read the text, attempt problems prior to each lecture.
8. Avoid talking to Professors.	Interact regularly with Professors outside the classroom.
9. Cut classes/don't get the most out of lectures.	Attend classes regularly and practice good listening skills. Ask relevant questions in the class and get your doubts clarified.
10. Fail to take notes; or fail to use the notes properly in the learning process.	Take notes properly; study the notes and understand the lessons.
11. Not reading the text-book properly and trying to complete the home work/problems without understanding the steps, procedures, or concept(s) involved.	Read and understand the general concepts before attempting to do the problems.
12. Fail to solve the assigned problems.	Solve not only the assigned problems but extra problems also. Understand clearly the lesson(s) taught.

Don't Hesitate to Seek Help

Who are the primary sources of "help" for your academic work?

- Your classmates and your friends.
- Your professors/teachers.
- Your mentor or proctor or students' counsellor.
- Sometimes, a member of your family, parents, or a relative might also help you in your academic work.

How Can You Get the Most in Your College Education?

Students who get the most out of their college education are those:

1. Who are more academically involved.
2. Who organize their study and time effectively.
3. Who have many interpersonal activities with their faculty members and their fellow students/classmates.
4. Who are the happiest in their class.

7

Importance of Reading in Your Learning Process

Reading of the class notes and textbooks is an important activity in the learning process. Learning involves many steps and stages:

1. Reading for comprehension and understanding of the subject matter.
2. By reading your class notes and textbook you begin to understand the theories, definitions, principles, methods, and procedures involved in the subject area and thereby gain confidence.
3. Reading helps you to solve the problems and to work out the solutions.
4. Reading helps you to take up the tests and examinations confidently.

Reading for Comprehension and for Understanding the Subject Matter

You should know:

- What to do before you read a textbook/journal paper.
- What to do while you are reading the text matter.
- What to do after you read the text matter.

Importance of Good Reading and Communication Skills

Just as you learn good behaviour from the examples set by your faculty/professors, you also learn to write and speak in good English by reading the best books and literature.

By reading the world's best books, you can:

- Dramatically improve your knowledge of vocabulary.
- Increase your comprehension skills.
- Learn to express yourself in an intelligent and concise manner.
- Improve your power of concentration.
- Enjoy exciting history/stories from different countries and at various periods of time.
- Read about the lives and deeds of the world's greatest people, scientists, engineers, and philosophers.
- Develop/stimulate your imagination.
- Improve your ability to think rationally and clearly.
- Learn to feel for others.
- Explore the good and evil of human nature, so that you can learn and distinguish the right from the wrong.
- Take part in discussions about the ideas and beliefs of the world's great thinkers.

These days, it is hard (difficult) to get into the habit of reading, as there are so many distractions while you are studying in the College. Even when books are read, some of them are of a mediocre standard, using a limited vocabulary, colloquial english, slag, incorrect grammar, poor examples of behaviour, as well as dull or depressing plots and shallow characters. Books like this do not teach better standards of english or encourage a higher level of thinking or improve communication skills.

Try to acquire a few of the world's great books and stories, and start reading them, you will slowly gain most, if not all, of the benefits listed above. The more you read, the better your english will become. Don't be scared of reading a book that seems a little difficult at first. It is only by pushing yourself beyond a comfortable limit that you will gain new vocabulary and improve your comprehension and communication skills.

A good way to really appreciate a book, especially one written many years ago, is to find out the approximate year (period) in which the book was written. This will give you an idea about the history of the time, the place, the scientific discoveries made at that time, the politics and philosophy of the era, the clothes being worn, the religion, the food habits and customs of the people/country the book is based on, and information about the author of the book. By doing this, you will understand the book in greater depth and add to your general knowledge as well.

Before You Read

- You should know the purpose of reading.
- There are two levels of reading:
  (a) Reading to get a general idea of the subject matter or to have an overview of the subject (This is called superficial reading or reading at the surface level, i.e., without going too deep into the subject matter).
  (b) Reading intensely to acquire an in-depth knowledge of the subject (i.e., reading for deeper understanding).
- Both levels of reading are equally important.
- Write marginal notes noting the important ideas as you read the text.
- Make or keep a list of questions and try to answer them after reading the particular topic in the subject.

While You Read

- Read actively (by concentrating your mind and not distracted by any other idea or thought).
- Try to understand the concepts thoroughly.
- Take your own time; don't try to read too fast.
- If there is any particular point or portions that you don't understand, make a note of it and get it clarified from your teacher or your friends.
- Periodically, stop and make a mental note of what you have read and understood.

After You Read

- Answers the questions you prepared before you started reading the text or notes.
- Read again, a particular portion, if required.
- Review everyday what you have read; Review on weekly basis.
- Solve the problems assigned/complete the assignment given by the teacher.
- Prepare for the test/exam by reviewing the class notes/ text and by solving additional problems (given at the end of the text).
- In some cases, you may have to understand the problem first before you attempt to solve it.
- Relate the problem to what you have read earlier and then try to solve the problem. Proceed in a logical way, by noting what are the parameters or values given and what is to be found out or solved.

How to Master the Text Information or Your Class Notes?

- Try to learn as much as possible from your lecture notes.
- Read the relevant text material and understand the information presented.
- Answer the questions given at the end of each chapter in the textbook.
- Solve the problems given by the teacher and the problems given at the end of each chapter in the relevant textbook.

How to Learn from Your Class Notes?

- Study and understand the information contained in your notes.
- Identify the important definitions, principles, rules or equations, and symbols the teacher has explained in the class.
- Redraw the sketch or diagram which the teacher has explained in the class.

- Try to write a summary of each lecture class. You can refer to the textbook but the summary should be written in your own words.
- Find the answers to the questions given in the Assignment/ Tutorial sheets.
- Reflect on (i.e., make a mental picture of) what you have learned.
- Review what you have learned on a weekly basis.

8

You should Develop Your Communication Skills

Your communication skills are important in your engineering profession.

- Employers want their employees to be good in their communication skills.
- Develop a positive attitude about your communication abilities; build on your strengths.
- Develop a plan of action to improve your communication skills still further.

Importance of Oral Communication Skills for an Engineering Professional

- In many instances, you may be required to talk and present about the progress in respect of the work assigned to you/your team.
- You may be asked to present your ideas/views on a new proposal/project.
- Will be useful in Project Review and Committee meetings
- In team collaborative work.
- Often needed during short courses and in technical training programmes.
- Required in professional society meetings/seminars.
- Useful in discussing with your subordinates about work related activities.

Seminar Presentations

Good planning and preparation can turn an ordinary power-point presentation (PPT) into an informative, lively, and interesting presentation.

Practical considerations demand that PPT should not exceed certain time frame (formal presentation for 20 minutes, and for Q & A session 10 minutes followed by PPT).

Considerations in Making the Presentation

- Of course, the contents: What do you want to share with the audience (facts, figures, findings of your study, your own perspectives, and experiences)?
- Understanding the audience (the learning purpose and what learning experience they should have).
- Maintain eye contact with the participants and speak to them and not to the screen.
- Ensure that the participants engage actively rather than jut receiving the information passively.
- Establish the relevance of the topic/seminar to the participants/audience.
- Give a background information and introduce the topic by displaying the title of the presentation.
- Explain the main concepts, your experiences and findings.
- Ensure that the illustrations/diagrams are placed at the appropriate places.
- Do not read your slides; paraphrase the contents.
- Use subtle transition effects when changing from one slide to next.
- Number the slides.
- Use short video/audio clips to reinforce the concepts visually.
- Supplement the PPT with a brief handout (for follow up and analysis).

- Go to the venue/seminar hall at least 15 minutes in advance to check projector, sound system, computer, and internet connection (if you are planning to use it).
- Get yourself familiar with the remote control/laser pointer (slide changer).

Note: that even the best presentations are quickly forgotten. Hence, make a positive impression of your presence in the seminar, and ensure that at least 1 or 2 important messages the participants carry with them to use and apply later.

Tips for Effective Power-Point Presentations

When preparing and using Power-point slides:

- Keep visuals simple—limit each to only one point.
- Use key words, not sentences.
- Use a simple lettering style.
- Use both capital and small letters.
- Follow the Rule of 6 × 6.
  - Use not more than 6 lines in one visual/slide.
  - Use not more than 6 words in one line.
- Keep lettering size such that when it is projected it is visible to the students in the last bench also.
- Face your audience and talk; do not speak to the screen.
- Use colours judiciously to add meaningful details and to ensure that the PPT slides are clearly visible/readable.
- Use a pointer/slides changer.

Listening Carefully

"Hearing is with the ears, listening is with the mind." It is not enough just to hear what some one else says, you listen with your mind as well. There are many good writers and speakers but few good listeners. Most of us filter the spoken words addressed to us so that we absorb only some of them—frequently those we want to hear. Listening is an art which not many people cultivate. But it is a very necessary one, because a good listener will gather more information and achieve better rapport with the other person. Both these effects of good listening are essential to good communication.

People do not listen because they are:

- Unable to concentrate, for whatever reason;
- Too preoccupied with themselves;
- Over-concerned with what they are going to say next;
- Uncertain about what they are listening to or why they are listening to it;
- Unable to follow the points or arguments made by the speaker;
- Simply not interested in what is being said.

The above reasons are called barriers of communication.

Good Listeners

- Concentrate on the speaker, following not only words but also body language which, through the use of eyes or gestures, often underlines meaning and give life to the message.
- Respond quickly to points made by the speaker.
- Ask questions, whenever required, to know the meaning and to give the speaker an opportunity to rephrase or underline a point.
- Comment on the points made by the speaker, without interrupting the flow, in order to understand and demonstrate that the speaker and the listener are still on the same wavelength. These comments may reflect back or summarize what the speaker has said, thus giving an opportunity for the speaker to reconsider or explain the point made.
- Make notes on the key points (this helps to concentrate the mind).
- Continuously evaluate the messages said to check whether they are understood clearly.
- Be alert at all times to the key message the speaker is saying.

- Allow the speaker to go on with the minimum of interruption.
- Be patient—it gives speakers time to think and explain their ideas and thoughts.

The 8% and 92% Syndrome in Spoken Communications

The syndrome refers to the messages that are not directly delivered, but emerge from between spoken lines. The American scientist Gerard Nierenberg refers this to the term which he has coined viz. "meta-message" (combined and higher message).

The spoken lines or words carry 8% of the verbal content of any communication, and every thing around that which is 92% is called "non-verbal communication". What constitutes non-verbal communication?

- The tone of the voice;
- The pitch;
- The rhythm;
- The gestures—face, body postures and hand gestures (body language); and
- The physiology of the speaker.

The above are the visual stimulus the receiver picks up unconsciously. They respond to the 92%, viz. what is called the meta-message, the packaging of the message.

How you say is more important than what you say. It means you have to be aware of your

- energetic response,
- the way you look,
- your body language,
- your breathing,
- what you feel inside at the moment, and
- whether your smile is genuine or fake.

Your meta-message should be more accepting, welcoming, and of friendly response. This will be immediately reflected in the other person in a positive way.

Body Language

Words are not the only means of communication. The way you present yourself and react physically can convey very effectively the extent to which you are really listening to what the speaker says. This is called body language. Some of the aspects of body language are:

1. Facing other people fully so that you are seen to be listening to what they have to say.
2. Maintaining good eye contact to demonstrate that you are listening—look the other person in the eye rather than somewhere over his shoulder. This eye contact should not be for more than 5 seconds at a time.
3. Maintaining an open posture to demonstrate that you are receptive to their ideas or comments. Do not be seeing or reading some papers lying in front of you.
4. Showing that you are interested by leaning slightly forward.
5. Smiling encouragingly from time to time.
6. Nodding your head in agreement when you want to convey that you are getting the message and are in agreement and happy with it.
7. If the person says something which you disagree, then remain neutral and tell orally that you have a different point of view.
8. Remaining relaxed and keep a good composure which will encourage an open and responsive relationship.
9. Maintain a distance of approximately 1 metre during the presentations.

Observing Body Language

Just as you can communicate through your own body language, you can help in the process of "getting the message" by observing the body language of others. Facial expressions and physical gestures (e.g., hand movements) will give some indication of the degree to which the person takes active interest in the discussion, listening to your comments/questions, agreeing or disagreeing with what you say, or angry, bored or indifferent.

Be aware, that sometimes body language may be deceptive (i.e., it may not be true).

Designating Roles in Groups (Group Discussion)

Groups that are created for in-class discussion can easily be organised following the four-person model. Each member of the group play a specific role that supports the team's collaborative effort. These roles are:

1. **Leader:** Responsible for keeping the group on task, maintaining the time schedule (meetings, deadlines), and maintaining contact information (phone number, e-mails).
2. **Facilitator:** Encourages conversation, prompts the members, and includes all ideas/opinion, and guides the discussion towards consensus building.
3. **Investigator:** Questions the members whether an idea are correct and ensures a solution is possible.
4. **Reporter:** Can be identified by the members of the group for the consolidation and for reporting/presenting the group work.

While some people will tend to lead and some will follow, everyone should participate and give their ideas in the interest of group unity.

Employers want that your Communication Skills should be Better and more Effective

A National Survey of over 5,000 engineering employers/ CEO's revealed that industry's concern is:

- Engineering institutions should give their students more instructions and practice in written and oral communications.
- This will also improve the employability of the students graduating from engineering colleges.

You should also Improve your Writing Skills to Become a Good Engineering Professional

Good writing skills are required:

- In drafting letters, memoranda, and e-mail correspondences.
- In developing design specifications.
- In writing project proposals.
- In writing your opinion/comment on a project proposal developed by your colleague(s) or group.
- In writing contracts, patents and other interdepartmental documents.
- In writing progress reports.
- In preparing good technical reports/articles.
- For writing and publishing papers in professional engineering journals.
- In making evaluation reports of your subordinates.

The Writing Process

Writing is a process that involves at least four distinct steps:

(a) Prewriting,

(b) Drafting,

(c) Revising, and

(d) Editing.

The above writing process is known as a recursive process. While revising, return to the prewriting stage to develop and expand the ideas.

(a) Prewriting

- Prewriting consists of points or ideas gathered before students write a draft of the document. It includes thinking, notes, talking, discussing with others, brainstorming, outlining, study in the library, assessing data.
- Although prewriting is the first activity to engage in, generating ideas is an activity that occurs throughout the writing process.

(b) **Drafting**

- Drafting occurs when ideas are put into sentences and paragraphs. Concentrate on explaining and supporting the ideas fully. Begin to connect the ideas. Regardless of how much thinking and planning is done, the process of putting ideas in words changes them; often the very words selected evoke additional ideas or implications.

(c) **Revising**

- Revision is the key to writing assignments. Which terms should be more explained or defined? Is the organisation of the document effective? Do the readers need to know certain prior information clearly so that they can understand this idea?
- At this stage the text is also further refined, made comprehensible and accurate as possible. The connections between the ideas are clear.

(d) **Editing**

- Check for correct grammar, sentences formation, and spelling. Carry out correction of spellings, put page no., and add a cover page and a contents page.
- Don't edit the writing until all the steps in the writing process are complete.

Develop a Strategic Plan to Improve your Communication Skills

- Take courses/instructions in oral and written communications.
- Look for opportunities to write (keep a record of your important activities briefly describing them); write an essay, article, poem or short story; send e-mails with correct spellings; (reading newspapers, magazines, technical journals and good English novels).
- Develop skill in writing Technical Reports/Articles.
- Look for opportunities to speak in English whenever possible (in Students' organisations, club, regular classes, professional society meetings).

9

Balance Your Study/Work and Recreation/Relaxation

- Strike a balance between your immediate and future satisfaction.
  - Too much immediate satisfaction—you don't get the work done; you feel guilty.
  - Too much delayed gratification—feelings of deprivation and resentment can destroy your commitment.
- Find a proper balance between work and recreation/relaxation.

How to Manage Your Stress

- **Eustress:** Positive form of stress; can motivate individuals to attain high level of performance and achievement.
- **Distress:** Negative form of stress, can distract you from getting the best.
- Attend training sessions on stress management.
- Learn strategies for coping up with stress and how to manage stress.

Group Discussion: Importance of Attitude

In your group, discuss the following quotation:

"The longer I live, the more I realise the impact of attitude on life. Attitude to me is more important than facts. It is more important than the past, than education, than money, than circumstances, than failures, than successes, than one's appearances, giftedness, or skill. It will make or break a life/person.

The remarkable thing is we have a choice everyday regarding the attitude we will embrace for that day. We cannot change our past. We cannot change the fact that people will always act in a certain way. We cannot change the inevitable. The only thing we can do is play the one string we have, our attitude."

"I am convinced that life is 10% what happens to me, and 90% how I react to it. And so it is with you. We are incharge of our Attitudes."

Select a leader amongst your group to lead the discussion on the above topic and keep a record of the important issues discussed and what you have learned.

Your Mental and Physical Health is also Important

- Follow tips for good health.
- Balance your work and play.
- Avoid stress and learn to manage your stress, both physical stress and mental stress.

Tips for Good Health

- Eat nutritious food; avoid fast foods (Junk foods).
- Engage in regular physical exercises/activities as far as possible.
- Get adequate sleep everyday.
- Do not take any medicine without Doctor's prescription/ advice.
- Avoid alcohol, drugs, tobacco, and smoking.
- Avoid over eating.

Benefits of Daily Physical Activity

- Reduces the risk of heart disease by improving blood circulation throughout the body.
- Keeps weight under control.
- Improves blood cholesterol levels.
- Prevents high blood pressure.
- Boosts energy level.

- Helps manage stress; releases tension.
- Improves self-image.
- Reduces anxiety and depression and increases enthusiasm and optimism.
- Increases muscle strength, increases the ability to do physical activities.
- Helps build and maintain healthy bones and joints.
- Provides a way to share an activity with family and friends.
- Establishes good healthy habits and helps to avoid obesity and other undesirable consequences that lead to heart attack and stroke later in life.
- Reduces the risk of developing diabetes, colon cancer, and many other diseases.
- Promotes physical and psychological well-being.
- Reduces the risk of dying prematurely.

10

Self-Study and Group Study

- You learn by yourself, i.e., self-study.
- You learn with others—It is called collaborative learning, as it occurs in a group study with one or two of your close friends.
- Identify by which method you learn maximum (You can get your doubts clarified when you study along with your friends, as in group study).
- Both the learning methods, self-study and group study are equally important.

Why do Students Study Alone?

Some of the answers given by students are:

- I learn more studying by myself.
- I can study at my own pace and convenience.
- I have my own learning style. Hence, I would like to study myself independently.
- I feel group study is a distraction for me/it is a waste of time.
- I learn very little when I study with my friend(s).
- I don't have anyone to study with.

Benefits of Group Study

However, there are some benefits of group study:

- You will be better prepared for the world of work when you join an engineering profession. In your work after graduation you may have to work with your colleagues and other engineers, wherein you may have to discuss and understand the job/task to be done by you and your team.

- You will learn more.
- Group study improves your interpersonal relationships and communication skills with others.
- Your friends can help you to clarify your doubts.
- You may enjoy group study more.

Self-Study and Group Study

- You should understand that both self-study and group study are equally useful and important.
- What is the ideal size of a study group? Normally it should not be more than 3 or 4 members (of your class).
- What can be done to keep the group from getting off the tasks? (from getting distracted!)

  The group should have an understanding that their main task is to study and learn the subject matter and get the doubts clarified.
- The duration of group study at a time should not be more than 2 to 3 hours in a day.

Study Groups: Study groups give opportunities to students to develop communication skills. They expose weaker students to peer coaching (i.e., learning from fellow students) and discussion that help them to succeed in the college, and they also allow participation to practice the team skills which are important in today's work environment.

College courses offered through team study (or study groups) give students opportunities to work together and learn together. Specially organised study groups will be particularly effective at providing opportunities for weaker students to learn better.

11

Learn to Manage Your Time

- Time is precious; you should learn how to manage your time.
- Time management will help you to accomplish maximum of your learning everyday and every week, in every semester.
- Schedule your time; i.e., allocate time for different activities of your daily routine.
- Devote definite time schedules for your daily activities like attending to:
  - (a) Personal needs;
  - (b) For studying in your room;
  - (c) Time for library study;
  - (d) Time for group study, if required;
  - (e) Recreation time with your friends;
  - (f) Time for games, sports, and other outdoor activities.
- Serious commitment to your scheduled study time is very important.

Benefits of Scheduling Your Study Time

- If you have a time schedule you wont feel you are stressed.
- Time schedule will help you to keep up with your studies.
- It will provide feedback as to how much you are studying and learning everyday or every week.

- Within the time period of a day or a week you will know what you can do and cannot do.
- If you follow a time schedule, it will give you a feeling that you have more time and that you are relaxed.
- If you adhere to your time schedule you will feel that you are not over-stressing yourself during your studies.

How Many Hours you Should Devote for Your Studies?

This is a tricky question. The answer depends on many factors:

- How difficult is the course?
- How good are you in your studies?
- Some are foundation or core subjects and some are prerequisite courses for learning other advanced subjects.
- What grade do you want to receive?
- During the first semester you may have to spend more time studying preliminary subjects like English, Mathematics, Physics, Chemistry, Engineering Drawing, and Computer basics; and you are expected to clear them, so that you can devote your time in the subsequent semesters in studying the core engineering subjects of your choice.

Making up Your Weekly Schedule

- Identify all your commitments, i.e., time for attending theory classes and laboratory classes, and time for travel, time for meals, etc.
- Balance time is available for two purposes:
  - For self-study, and
  - For your recreation (extra curricular activities, games, etc.).
- Schedule/apportion your study time to avoid wasting time.
- What should I study and review on a weekly basis?
- Where should I study?

- What subject(s) should I study?
- How much time I devote for studying each subject?

These are the questions which you must answer so that you can arrive at a workable time schedule.

Seven Tips for Time Management

What you accomplish during a 24-hours period depends on your own motivation, your energy, your skills, talents and abilities, and other resources. Time management is not a way to make you work harder and longer, but a means to help you work smarter to accomplish your work more easily and rapidly.

Here are some strategies you can use to overcome the obstacles that are affecting your time.

1. Set clear goals for yourself: You will want to accomplish many things in your life. You must be able to determine what those effective goals are and make sure your efforts are always directed toward their achievement. Effective goals should be: Specific, Measurable, Achievable, Realistic, and Time-based (SMART).

2. Prioritise: It is wise to periodically make a list of the tasks that confront you and prioritise them. Determine from the list the things you think are most important to accomplish. A good leader knows when to concentrate on the important and eliminate the rest.

3. Organise: As you prioritise tasks and set deadlines, you will want to plan and carry out the activities. Use a planner with a calendar with enough space to make notes for daily activities, contact information, and "to do" lists. You will find this to be an indispensable tool for managing your time. Using a computer also helps in organising your time.

4. Being able to say "No" when it is appropriate: As a leader, you will find that you are called on to do much more than you have been asked to do before. When you learn to say "No", you are not closing the door on your responsibilities, but rather you are making sure that you can meet your commitments and accomplish the maximum possible in the time available to you.

5. Concentrate on the current task: Concentration can be difficult when you have a lot on your mind. Your time will be better spent if you are able to focus on your goal and avoid interruptions.

6. Use your "personal prime time": Everyone is different. When you plan your tasks, think about your own "prime time"—the time you can concentrate on the more challenging tasks when you are at your best.

7. Celebrate: Celebrate the achievement of your goals to maintain a healthy balance in life between work and play. Reward yourself when you complete a task. If you have worked in a team by delegating some tasks to others, reward the efforts of all involved.

Develop a plan to fit your own needs—be flexible to some extent as your needs change.

Procrastination

Procrastination is the habit of postponing the work that we know should be done now itself, but instead we engage ourselves in some other activity.

Avoid Procrastination

Reasons for Procrastination are:

- Fear of failure (i.e., a feeling that work may not be done successfully).
- Fear of success (i.e., the work is done successfully but then what to do next?).
- The work may involve some unpleasant task, which one may not like to do.
- The individual is disorganised (i.e., he does not know what to do?).

A major obstacle that can prevent you from practicing good time management skill is procrastination.

Procrastination is continuous pushing aside of the tasks and postponing the work that needs to be completed.

Some ways to overcome Procrastination are:

1. Make the task meaningful: Ask yourself why the task is important to you and how it relates to your long-term goals. We often lose sight of our goals. A reminder of the goals is a good idea, especially when you are affected by procrastination.

2. Break the task into its elements: Breaking large assignments into manageable parts will always help. Set dates/targets to work on each of the elements or steps. Once you have started, the task will not be seen so hard or difficult.

3. Keep yourself organised: A lot of time can be saved by having the right materials/information with you when you are ready to work on a task or assignment.

4. Be positive: Avoid speaking negatively about the task. Be positive and have confidence in your ability to complete the task. Tell yourself, "I know that I can finish this work and I am capable of it".

5. Plan, do it and complete the task: The moment you find yourself procrastinating, try to complete the task.

6. After completing the job, praise yourself: Think of the task that you have accomplished.

Manage your time, even if it appears that you have all the time in the world. Stop procrastination before it begins!

12

Productive Actions, Positive Thoughts, and Positive Feelings

There are Two Types of Actions

1. **Productive actions:** Actions which support the achievement of your goals, aims, and aspirations.
2. **Non-productive actions:** Actions which interfere with your profession, work or studies against the achievement of your goals, aims, and aspirations.

Obstacles Faced in Choosing Productive Actions

- You have difficulty in choosing to do things you don't find enjoyable/easy to implement.
- You are afraid to take productive action because you think that if you take the productive action and still fail, it will reflect on your capability and you may lose the confidence.
- You prefer to blame your failures on other people or other factors external to yourself.

Value Judgment Applied to Your Thoughts

There are two types of thoughts:

- **Positive thoughts:** Result in your choosing productive actions and the right actions.
- **Negative thoughts:** Result in your choosing non-productive actions, wrong actions or not doing things you are expected or supposed to do or act upon.

There are Two Types of Feelings

- **Positive Feelings:** Produce positive thoughts, which in turn lead to productive actions and desirable behaviours.
- **Negative Feelings:** Produce negative thoughts, which in turn lead to non-productive actions; wasteful actions or not doing activities you are supposed to engage in.

How to Make Behaviour Modification Work for You?

Behaviour Modification

Choose productive behaviours and work to transform negative thoughts to positive thoughts; and good feelings and good behaviours will follow.

For realising success you must follow the three steps strategy:

Step 1: Acquire Knowledge: "You know what to do".

Step 2: Have Commitment: "You want to do it" and "You have a desire to do it".

Step 3: Implement New Thoughts or Ideas: "You actually do it".

Formula for Success

According to Shiv Khera, the famous and well known motivation speaker and author, your attitude, you values, and your vision form the basis of your success. Anything is possible to achieve by motivating yourself with your positive thinking.

However, positive thinking alone, cannot guarantee success. Positive thinking, coupled with a well defined action plan, will yield success.

One's willpower more than skill contributes to one's success. Even though technical skill (hard skill) and people skill (soft skill) are essential, it is people skill that really matters and is of vital importance as you climb higher in your professional career.

By building cordial relationships many problems can be solved. If people skill is lacking you cannot taste success.

Positive thinking and positive behaviour should become an integral part of one's life. Developing good habits are essential, for it greatly contributes to positive attitudes and positive thinking. Give importance to your reputation and most important to your character. Keep up your self-esteem.

Don't quit. Don't give up.

—Abraham Maslow
(1908-1970)

Maslow's Theory of Motivation

Maslow's Hierarchy of Needs identifies five levels of needs, which are best seen as an hierarchy with the most basic need emerging first and the most ultimate need last. People move up the hierarchy one level at a time. Fulfilled needs lose their importance and the next level of needs are activated and strengthened.

1. **Physiological needs**: Food, water, air, and shelter.
2. **Safety needs**: Security, orderliness, and freedom from fear.
3. **Belongingness**: Feeling of family and friends.
4. **Esteem needs**: Self-respect, achievement, fame, and reputation.
5. **Self-Actualisation**: To become aware of what you are most fitted for—understanding fully your potentials and working hard to realise your full potentials.

What are Your "Needs" and "Wants"?

1. *Needs* are things that you must have, things that are essential.
2. *Wants* are things that you desire to have.

What is Self-esteem?

Self-esteem is appreciating one's own worth and importance and having the strength and character to be accountable for one-self and to act responsibly towards others.

Self-esteem is made up of two components:

1. **Self-capability**: Sense of competence; ability to produce desired effect.

2. **Self-respect:** Sense of personal worth and importance.

Level I: Physiological needs are the most basic human needs. They include food, water, and comfort. Physiological needs are satisfied by a regular monthly income.

Level II: Safety needs are the desires for security and stability, to feel safe from any harm. Safety needs are satisfied by having a secured job and a home.

Level III: Social needs are the desires for affiliation. They include friendship and belongingness to the organisation and family group. Social needs are satisfied through family get-together, games, sports, parties, and celebrations. Social needs are fulfilled by showing concern and care.

Level IV: Esteem needs are the desires for self-respect and respect or recognition from others. Esteem needs are fulfilled by matching the skills and abilities of the person with the job and by appreciating the work done by the person.

Level V: Self-actualisation needs are the desire for self-fulfillment and the realisation of the individual's full potential. Self-actualisation needs are fulfilled by tasks that challenge one's minds consistent with his/her aptitude and training.

Motivation

Motivation of individuals involves:

- Understanding their needs—Are they interested primarily in job security, status, advancement in their job/profession/career, achievement or money?
- Remembering that money is not the only reward that most people need or want. People can also be motivated by recognition, praise, and opportunity to make the best use of their knowledge, abilities, skills, and to enhance them.
- Considering the importance of expectations as an influence on motivation, a reward whether financial or non-financial will be much more effective when people know what they can get if they work well enough.

- Creating conditions where individuals can best satisfy their own needs by directing their efforts towards achieving the success of their team, department and, ultimately, the organization. This is why one of the arts of leadership is to get people to think for themselves about what they can and should do and by getting them to understand and agree to the standards, norms, and targets they are expected to achieve.
- Recognising the fact that people can be motivated by the work itself if it satisfies their need for responsibility and achievement. This can be done by:
  1. Increasing individuals responsibility by "empowering" them;
  2. Giving them more scope and freedom to vary the methods and sequence (wherever it is possible, consistent with the general practices), and pace of work;
  3. Giving people a complete unit of work (wherever possible); and
  4. Giving people the general expectations, to monitor their own performance.

Understanding Others and Respecting the Individual Differences

- There are differences in personality types and in the learning styles of individuals.
- There are cultural, regional, and gender differences.
- You may have to improve your performance in cross-cultural environment, i.e., by communicating and working effectively in a group where people from different cultures, regions, and genders work together.

You Should Draw up a Personal Development Plan for Yourself

This means you have to

- Think and create a personal development plan, i.e., an action plan for improving yourself.

- Identify areas for improvement (you should be receptive to changes, strive to change, grow, develop and improve yourself continuously in every areas that impacts your work, study, life, and performance effectiveness).
- Your personal development depends on area(s) in which you want to grow, change, or develop to achieve your goals.
- Prioritise them in the order of importance.
- Work on them consistently and continuously.
- In addition, develop knowledge, skills, and right attitudes expected of a professional.
- Take part in as many co-curricular and extra-curricular activities as possible.

13

Your Personal Growth and Development During Your Studies

- You should be receptive to changes. Strive to change, grow and improve yourself continuously in every area that impacts your study, life, and performance effectiveness.
- Your personal development depends on area(s) in which you want to grow, change, or develop to achieve your goals.
- Major goal being receiving your B.E./B.Tech Degree in the chosen field of engineering at the end of your studies in the engineering college.
- In addition, develop knowledge, skills and right attitudes expected from an engineer working in the company, industry, society, or in your professional life later.
- Take part in as many co-curricular and extra-curricular activities as possible while you are studying in the engineering college.

Co-curricular and Extra-curricular Activities

- Taking part in co-curricular/extra-curricular activities will help you to broaden the scope of your education in the engineering college.
- No one will make you to take part in co-curricular and extra-curricular activities since all these activities are undertaken by the students voluntarily.

- These activities will have greater educational value than even your coursework requirements.

Co-curricular Activities

These are:

- Participation in students' organisations/clubs.
- Participation in engineering projects organised by Institute/College.
- Summer training in an industry, company, or business enterprise.
- Taking part in social service activities, community service, red-cross activities, road safety week, tobacco/drug awareness camp, NSS (National Service Scheme) activities, blood donation camp, etc.

Benefits of Participation in Engineering Projects at Institutional Level or Departmental Level

- Students' design competitions are held in engineering colleges and technical institutions. It is important that you take part in these events, as it will add value to your engineering education at the college.
- Take part in technical paper contests/paper presentations.
- Design seminars/design contests.
- Take part in undergraduate research. This will help you to know how research project work is carried out in an institutional set-up/R&D organizations. You may be awarded extra credits for this work. This also helps you to get an exposure in engineering R&D work.

Benefits of Industrial Training During Your Engineering Studies

- This helps you to develop job related skills.
- You can gain exposure to engineering practices/industry standard practices and field practices or business practices.
- Apply your knowledge, skills, and abilities you have acquired during your studies to practical situations and

thereby get work experience which will be useful for your future employment.

- Strengthen your motivation to succeed in you studies.
- You can get good references (recommendation letters) for future employment.

Types of Students' Organisations

- Students' chapters of various professional societies.
- Participation in literary clubs/associations.
- Take part in Students' Council activities.
- Organising and taking part in students' clubs such as:
  - Entrepreneurship club
  - Dance club
  - Dramatics club
  - Fine-arts club
  - Health club
  - Hobby club
  - Movie club
  - Karate club
  - Music club
  - Quiz and G.K. club
  - Out-door/trekking club
  - Astronomy and space club
  - Photography club
  - Creativity club
  - Toast-master club (for promoting English speaking skill).

What are the Benefits of Participation in Students' Organisations?

- It meets your social needs.
- It develops your leadership skills and organisational abilities.

- It engages you in technical and professional development activities.
- It improves your personality traits, communication skills, and interpersonal relationship.
- Able to participate in students service activities.

14

Academic Integrity and Ethics

Students are expected to attend the classes as scheduled and be prompt in submitting assignments and tutorials. They should not use inappropriate languages in the classroom, offer unwanted remarks, or jokes in class or in the meetings. Students should fulfill in a reasonable way the requirements and expectations of the courses as stated/required by the faculty.

Plagiarism (Copying from Other's Work)

Plagiarism means "Ideas, words, or statements of another person as one's own without appropriately quoting the source." In assignments/projects you are required to give credit for the ideas of others making a specific point of what you have done and how you have proceeded further with that idea.

Students must perform their work diligently and work ethically. They "must not copy, steal, change, destroy, or impede another student's work".

Plagiarism is defined as to steal the ideas or thoughts or words or language of another person and pass it off as one's own original work. Plagiarism means copying from other's work. However, you may need to use or refer other people's work to support your findings and arguments in your research paper or in your projects. You can use information, ideas, work or data from other source but give the reference, viz. details of

the journal, author's name and title of the paper from which the idea is taken for your work.

Tips for avoiding Plagiarism are:

1. Explain/describe the idea(s) from the original source in your own words; mention the title of the paper, name of the another(s), Vol. No. and the page numbers (pp).
2. Document any information or ideas taken from other sources, textbook, report, journal, paper(s), etc.
3. Give credit to and acknowledge the original source by quoting the author's name(s), title of the paper, name of the journal in which the paper has been published, Vol. No., Issue No., and the page numbers (pp).
4. If you are paraphrasing, use a sentence that credits the original source somewhere in the paraphrase. Do not indulge in plagiarism since electronic software have been developed to detect any form of plagiarism when you submit paper(s) for any conference/seminar or journals.

Basic rules and regulations of the Institution/Library should be followed.

Students must not rip (or tear) from journals and books kept in the libraries. They also should not deface any page or portion of the book.

Violations of Academic Integrity

Academic fraud and dishonesty includes plagiarism, copyright infringement, cheating, fabrication, multiple submission of assignments, obtaining unfair advantage, unauthorised access to academic or administrative systems/records, aiding and abetting, impersonation and threatening to harm others.

1. **Cheating:** Using unauthorised notes, aids, or information in an examination; making alterations in the assignments already graded for any undue advantage. Students should not give his/her own work/assignment to their friends/classmates to submit it for grading.
2. **Fabrication:** Giving false information, data, or citation; failing to provide an accurate account of

how information, data or citations were gathered; altering documents affecting academic records; forging signatures or making alterations on official academic document, grade sheet, letter, form, ID card; submitting false certificate; giving false excuses for absence, delay, or illness.

3. **Multiple Submissions:** Submitting identical papers/ assignments or course work for credit in more than one course.
4. **Obtaining Unfair Advantage**
   - Gaining access to examination materials prior to the time authorised by the teacher.
   - Stealing, defacing, or destroying library or research materials.
   - Unauthorised collaboration on an academic assignment.
   - Retaining, possessing, or circulating previously used examination materials without the faculty's permission.
   - Obstructing or interfering with another student's academic work.
   - Offering bribe to staff or any university/college employee to effect a grade change, or to gain unfair advantage over other students.
5. **Unauthorised Access:** Viewing or altering in any way computer records, modifying computer programmes or systems, releasing or distributing information gathered via unauthorised access, or in any way interfering with the use or availability of computer systems/information.
6. **Aiding and Abetting:** Providing material, information, or other assistance, which violates the standards for academic integrity; providing false information in connection with any inquiry.
7. **Impersonation:** Impersonating or allowing to be impersonated by another individual during classes, examination, or in other university activities.

8. **Threatening to Harm:** Threatening students, doing bodily harm or causing financial loss to any student, faculty, staff or administrator.
9. **Misconduct:** Behaving in a manner that violates or adversely affects the rights of other students or members of the academic community (disrupting class/campus activities, unruly behaviour, etc.).
10. **Copyrights Infringement:** Using copyright materials (print material, non-print material, software, electronic, or multimedia) in a manner that violates copyright laws.
11. Students should not indulge in any theft or damage of the university/college property.

Self-plagiarism

Self-plagiarism, referred to as "multiple submissions". Self-plagiarism involves copying from one's own work without proper attribution/reason. Copying one's work is dishonest because it presents the material as original work. It's also unfair since students who don't have a previous report to rely on have to do a lot more work to earn the same grade or marks. More important, students who simply recopy their previous report are also cheating themselves of an opportunity to learn more.

Many students don't think plagiarising is wrong, more so with self-plagiarism. Students should be told that writing/copying from their previous reports is a form of cheating.

We can give students a scholarly alternative to self-plagiarism. Have them treat their previous course project/report as a publication and then build on it. They can appropriately cite their previous report as a reference and use quotations if they quote from it. A copy of the previous report can be appended to the new report.

Students shouldn't copy from reports they have written for other classes or overlapping projects.

We can control self-plagiarism by first identifying the problem, setting appropriate rules/guidelines, explaining procedures to our students, providing a scholarly alternative, modeling appropriate behaviour and occasionally checking for compliance. A more uniform playing field will be the result with more balanced student workload and increased student learning.

Ethics

Ethics means right conduct, following righteous path, moral codes, observe standards and norms of society.

Personal ethics consists of honesty, integrity, truthfulness, fairness, etc., e.g., giving bribe is an unethical practice.

Ethics deals with concern for others. To be ethical means to be truthful, kind, supportive, helpful, sympathetic, fair, and encouraging.

Self-interest should not drive individuals to use unfair means to achieve his/her goals. Being ethical means not to steal, bluff, conspire, criticise, betray, hurt, lie, manipulate, falsely accuse, and spread rumors.

Code of Ethics of Engineers*

The Fundamental Principles: Engineers uphold and advance the integrity, honour, and dignity of the engineering profession by

1. Using their knowledge and skill for the enhancement of human welfare;
2. Being honest and impartial, and serving with fidelity (faithfully) to the public, their employers, and clients;
3. Striving to increase the competence and prestige of the engineering profession; and
4. Supporting the professional and technical societies of their disciplines.

The Fundamental Canons (Generally accepted Rules)

1. Engineers shall hold paramount the safety, health, and welfare of the public in the performance of their professional duties.

* Accreditation Board for Engineering and Technology; New York, U.S.A. (ABET).

2. Engineers shall perform services only in the areas of their competence.
3. Engineers shall issue public statements only in an objective and truthful manner.
4. Engineers shall act in professional matters for each employer or client as faithful agents or trustees, and shall avoid conflicts of interest.
5. Engineers shall build their professional reputation on the merit of their services and shall not compete unfairly with others.
6. Engineers shall act in such a manner as to uphold and enhance the honour, dignity, and integrity of the profession.
7. Engineers shall continue their professional development throughout their careers and shall provide opportunities for the professional development of those engineers under their supervision.

15

Preparing for Tests/Examinations

- Schedule your time for review of lessons you have studied prior to the tests/examinations.
- Revise the portions from your class notes/textbook(s).
- Develop practice to perform under time pressure (In some cases, within a short time you may be required to prepare for the examination and do some important work following certain definite guidelines/steps).
- Get a good night's sleep before your test/examination.
- Make sure you have the right materials and other items for appearing for the tests/examination.
- Arrive early for the test/examination.

Test-Taking Strategies

- Read/glance the entire question paper so that you have an overall idea of the test/examination.
- Answer the easier questions first.
- Work out the easier problems first.
- Be aware of the time available for writing the test/ examination.
- Allocate the required time for each question, so that you are able to answer all the questions; otherwise you

may not have time to write the answers even for easy questions.

- Try to complete a problem before leaving it; at the same time be conscious of the time you are taking to solve this particular problem.
- Allocate 5 or 10 minutes to check and recheck your answers (never leave a test early).
- Try to write legibly and neatly and don't be careless in answering the questions.

Verbs (Words) used in Question Papers and their Meanings

Following are the various directive verbs or words used in the Examination Question Papers. You should know the meaning of these words so that you can answer in the most appropriate way depending on the scope and nature of the subject.

Analyse	Show the essence of it by breaking it down into its component/constituent parts and examining each part in details.
Argue	Present the case for and against a particular statement.
Compare	Look for similarities and differences between the given statements.
Contrast	Discuss the two ideas which are in opposition in order to show the differences between them.
Criticise	Give your judgment about the merits or opinions about the ideas stated, and support your judgment by discussion.
Define	Give the definition as universally stated and accepted. Give the precise meaning of the word, phrase, or term.
Describe	Give a detailed information or description giving the important characteristics of the item.
Differentiate/ Distinguish	Show the difference(s) between the two items stated.

Discuss	Investigate or examine by giving arguments, analyse, giving reasons for and against.
Enumerate	Specify and describe the salient points/features one by one.
Evaluate	Make an appraisal of the value of the statement or the idea of its truth or use. Give your personal opinion, supported by evidence.
Examine	Check the validity of the statement and comment on it.
Explain	Make it clear, interpret, and account for giving details.
Explore	Thoroughly enquire by examining and analysing every aspect of it.
Illustrate	Explain clearly by giving concrete examples with figures, diagrams, or sketches.
Interpret	Bring out the meaning of it clearly and explicitly; also give your own view of it.
Investigate	Search, study, and critically analyse all aspects of the topic/problem.
Justify	Defend the statement or show adequate proof for your decisions or conclusions.
Outline	Give the main points and general principles of the topic omitting minor details, emphasising structure, sequence, and relationship.
Prove	Demonstrate whether it is true or false by presenting evidence; establish the relationship based on first principles.
Relate	Narrate or show how things are connected to each other and to what extent they are alike or interrelated with each other.
Review	Make a survey examining the subject critically.
State	Give the precise meaning/definition of the term. Specify details fully and clearly.

Sketch	Draw an outline diagram, drawing or figure of the equipment, circuit, or process.
Summarise	Give a concise account or the important points of the topics, omitting details and examples.
Trace	Identify the important milestones/events and describe the development or history of a topic from the beginning or from a particular period/ specific instance.

16

Employment Opportunities and Job Search

How Do You Identify Your Employment Opportunities?

- Look for announcements by your placement and training department/division for potential campus interviews.
- Attend on-campus interviews.
- Use the internet and apply on-line.
- Check other sources of employment.
- Consult your professors, teachers, and seniors to help you in your job search.
- Read "opportunities" column/section published by newspapers.

How Do You Conduct a Job Search?

- Prepare a resume/curriculum vitae (your certificates and other credentials).
- Prepare a covering letter.
- Develop your interviewing skills.
- Identify various employment opportunities available.
- Contact your placement and training officer/department for counselling and guidance.

The 21st Century Skills

While you are studying in the engineering college, it is expected that you learn and acquire 21st century skills, in addition to acquiring technical competencies in you chosen professional field. These skills are:

- Learning to know,
- Learning to live together,
- Learning to do, and
- Learning to be.

The 21st century learning encompasses the skills that are significant and important for you to acquire.

- **Collaboration:** The ability to work in cross functional team(s).
- **Critical thinking:** Taking on complex problem.
- **Oral communications:** Presenting the information.
- **Written communication:** Through writing assignments.
- **Technology:** Use of information and communication technology (ICT).
- **Learn about careers:** Through internships and industrial training.
- **Citizenship:** Take on civil and global issues; and learning through active community service.

Professional Skills

You are expected to possess and continuously develop the following professional skills.

1.	Introducing yourself (creating a positive impression).
2.	Listening—understanding what people say.
3.	Listening—showing interest in people.
4.	Communicating your feelings (positively).
5.	Dealing with anger/hostility.
6.	Responding to praise.
7.	Responding to expression of anxiety.

(Contd...)

8.	Responding to negative feedback.
9.	Coping with apathy and expressions of disinterest.
10.	Coping with silence during conversations.
11.	Appreciating other people's feelings.
12.	Giving information correctly.
13.	Simplifying complex information.
14.	Advising on emotional issues/difficulties.
15.	Seeking clarification.
16.	Asking open-ended questions.
17.	Waiting for replies.
18.	Changing direction or trend of conversation.
19.	Expressing support to others.
20.	Self-disclosure of personal information as appropriately required.
21.	Making a conversation more serious.
22.	Making a conversation less serious.
23.	Summarising what people have said.
24.	Holding someone's interest and attention.
25.	Ending conversation in a positive way.
26.	Organising skills for carrying out a given task.
27.	Skills in time management.
28.	Liaising with others.
29.	Resolving conflicts and misunderstanding.
30.	Developing self-confidence.
31.	Making a conversation a pleasant experience.

Skills Set Employers are Looking for While Recruiting their Employees

- Strong technical skills (hard skills).
- Critical thinking skill.
- Ability to use knowledge to solve real life problems.
- Soft skills (such as communication skills, interpersonal skills, and ability to work in diverse groups).

- Ability or willingness to work hard.
- Creative problem-solving (creative thinking skill).
- Curiosity.
- Professionalism.
- Resourcefulness (whether the individual is enterprising?).

Critical Thinking

Critical thinking is the disciplined mental activity of evaluating arguments or propositions and making judgements that can guide the development of one's beliefs and taking action. Evaluation of an idea is considered equivalent to critical thinking and it focuses on making assessment or judgement based on an analysis of a statement or proposition. Learning the process of critical thinking is best facilitated by a combination of didactic (by telling) instruction and experience in specific areas.

Techniques that are often taught as part of critical thinking process are generally more linear and serial, more structured, more rational, analytical, logical, objective, and goal oriented. (It sometimes referred to as left brain thinking or activity.)

Creative Thinking

Creative thinking requires an individual to look at parts and relationships and then to put these together in a new and novel way, as well as looking at the elements and the whole in a new perspective altogether.

Techniques taught as part of creative thinking process tend to be holistic, non-linear, parallel, more emotional and intuitive, subjective, more creative, more visual and kinesthetic. (It sometimes referred to as right-brain thinking or activity.)

Core Competencies (Soft Skills) Employers Look for in Recruiting Engineers and Managers

- **Innovation:** Capable of generating new and unique ideas.
- **Decision-making:** Selects the best course of action from amongst several feasible alternatives.

- **Self-awareness and Self-assessment:** Ability to learn about one self and recognise one's own strengths and weaknesses.
- **Drive and Enthusiasm:** Highly motivated bubbling with zeal and interest.
- **Interpersonal Relationships:** Ability to express and communicate so as to maintain cordial relationship.
- **Strategic Thinking:** Developing a creative solution combining own ideas with others.
- **Resilience and Endurance:** Ability to work under pressure.
- **Valuing others Ideas:** Making certain that others get their due credit for their ideas.
- **Collaboration with Others:** Promoting team work and team spirit.

Decision-making

The ability to make smart choices or decisions is a fundamental professional and personal skill of an engineer. With compressed project development times, decisions must be made quickly and prudently.

All factors relevant to a decision need to be considered and judiciously evaluated.

Decision-making is simple if you know the old maxim: "Divide" and "Conquer". Break your decision into the elements, think about the elements, and combine your thoughts to make up your mind. This makes tough decisions more manageable.

There are eight key elements for making smart choices and decisions. Focus on one element at a time. Do your own thinking first, before consulting others about alternatives. Give your mind a complete freedom.

(1) **Define your Problem (Problem Definition):** Define your problem clearly and unambiguously. (In many instances the problem/issue itself would not have been properly identified or known.)

(2) **State the Objectives:** You need to understand your objectives clearly. Clarify what you are trying to achieve with your decision. This is a key step in the decision-making process.

(3) **Consider all Alternatives:** Consider all alternatives (possible alternatives). Is there any constraint that limits your alternatives. Some constraints may be imaginary.

Some alternatives may be worst alternatives; may not be acceptable to all; and difficult to implement. Some may be feasible and better alternatives. Arrive at the best possible solution.

(4) **What are the Consequences?:** Make a judgement as to how well each alternative meets your objectives. Choose the best alternative.

(5) **What Compromises you can make?:** Consider what compromises you can make if you can't achieve all your objectives (in the beginning itself). Limit your objectives, if required.

(6) **Consider Uncertainties:** Identify and quantify (if possible) the major uncertainties affecting your decision. (What is the worst possible outcome of the decision you have made?)

(7) **Risk Tolerance:** How much risk you can take in implementing the decision you have arrived at?

(8) **Link Decision (Will it lead to further decision-making process?):** Identify and plan ahead how you can effectively coordinate the future decision, if desired. (Will it lead to another problem/linked decision?)

Decision-making is like any other skill, which can be mastered by practice. Understand each element of the decision-making process and systematically apply available data and your own professional judgement.

How do Your Employers Measure Your Potential When They Interview You for a Job?

- Your academic performances in the school and 10 + 2 classes.

- Your academic performance in the engineering college.
- Your achievements in co-curricular and extra-curricular activities.
- Your personal qualities and talents.
- Your references/letters of recommendation from your professors, teachers, and company executives under whom you did project work/industrial training.
- Your performance during the interview.
- Your attitude towards your work.

Frequent Questions Asked by Employers While They are Recruiting Their Employees/Graduates

Among the applicants of equal ability, what any successful job interview ultimately comes down to is giving the right answer(s) to the questions posed by the recruiter. The key questions that, in one form or another, almost every good interviewer asks are:

1. **Tell me/us about yourself (or can you introduce yourself?)** (Here you talk about your background, educational qualifications, place/institutions studied, work experience, if any, and your career ambition.)
2. **Why have you applied for this job/company?** (If you have left your previous job, if any, tell directly quickly and straight forward why you have left your previous job: "The job they put me in wasn't the one to my liking or it has not suited me:" then tell why?)
3. **What can you do for us?** (For this you should know about company/organization in which you are seeking a job. This indicates that you have done some spadework before the interview. Tell them in which functional area you would like to work and how will you contribute to the company; how can the company be benefited from your knowledge and experience.)
4. **What are your strengths? (or make a SWOT: strength, weakness, opportunity and threat analysis of yourself)** (Of course, your qualifications, knowledge, and experience,

if any, are your strengths on which you should build up. In addition, you should be able to list some of your other strengths, which you may have, such as energy level, enthusiasm, interest in the job, assertiveness, hard working, computer skills, communication skills, languages you know, adaptability, and flexibility.)

5. **What are your weaknesses?** (Identify some of the weaknesses you may have; you may not have all the strengths/attributes listed above; identify some of them and tell how are you planning to overcome them.)
6. **What are your significant accomplishments/ achievements?** (Give your achievements such as awards (academic, sport, etc.), Prizes won, any social service work done/community service done, organised any seminar/workshop/conference, college club activities done, any publications, creative work/innovation done.)
7. **What emoluments/salary are you looking for? (Your expectations?)** (Do not underestimate your potential; don't tell a low figure which shows your lack of confidence; also do not ask for very high salary; be realistic; give a range of your expectations and any specific benefit you are looking for.)
8. **Is there anything you would like to ask us?** (Utilise this opportunity to ask them to tell about your job contents, department in which you will be working, place of work, or any specific work/plan they may have for you.)

Preparations before job interview always help in successful interviews.

17

Creative Process and the Creative Mind

Creativity is connected to innovation and invention. Creativity cannot be easily defined. Creativity is not logical, and is not a process that comes about in an orderly, sequential, and step-by-step fashion. Many creative thinkers have found success by being forced to work outside conventional norms.

Creativity is "thinking outside the box" (i.e., not in a stereotyped manner with fixed ideas). Inventive scientists, engineers, and professionals have the unique ability to look at a set of problems and come up with a solution that no one else would have seen or imagined. The No. 1 trait among creative thinkers is their perseverance in solving a problem and arriving at a solution. Creativity is an innate (i.e., inborn) ability. The other factor in the inventive process is the ability to work in collaboration with others and with diligence and hardwork (i.e., doing the work thoroughly).

Creativity must be tempered with critical thinking, i.e., balancing analysis with innovation.

There are many problem-solving techniques which creative people are aware of; and hence are better of problem-solving.

How Creative/Inventive Minds Work? (Creative Process)

There are five stages of the creative process.

First stage is the arrival of the idea (ideation). The next four steps involved gradually in making that idea public.

Second stage is to expose the idea to criticism by close friends and colleagues.

Third stage is analysis, reinforcing with more information and details to a plan.

Fourth stage is learning to teach the idea to others (i.e., sharing the idea with others).

Finally in the **Fifth stage,** the inventor would take the idea to the public.

Characteristics of creative people:

1. Strong desire to change the status-quo.
2. Perseverance in finding a solution to a problem (Most of the ideas that look good on paper may not really work in reality/practice, hence perseverance is important.)
3. Ability to work in collaboration with others (teamwork), diligence, and hardwork.
4. Have a risk taking tendency. (They recognise that failure is part of the creative process.)
5. Have the attitude of not just searching for one right answer but arrive at multiple answers/solutions.
6. Have inventive minds, high degree of autonomy, superior general intelligence, a somewhat detached attitude in interpersonal relationships, and high degree of control over their personal environment.
7. Surround themselves with other creative people. (Socialization is also an important part of creative process.) They have a sense of social involvement.

What Needs to be Done to Improve Creativity Amongst the Students/Learners?

1. Doing exercises in enhancing cognitive abilities, such as free association and divergent thinking (lateral thinking) enabling students to think in different alternative ways.
2. Exposing the student to problem-solving techniques, such as flow charts, fish-bone diagram, etc. and segmenting the innovation process into smaller parts,

finding individual solutions, then integrating them as a whole.

3. Assigning projects that encourage open-ended problems solving. This improves creative behaviour amongst students/learners.
4. Encouraging students to make minor or routine improvements to existing systems. (Inventions that are found on new principles to perform the primary function of an existing system are also creative processes.)
5. Working in teams is an essential part of the creative process, because it allows students to expand their minds and explore different solutions to problems. A collaborative environment energises the creative process/creative cycle.

Focussing on innovation and invention is the need of the hour. Curriculum designers should make a note of this.

A good knowledge based education coupled with its application in diverse fields and analysis will sharpen the creative mind and the creative process.

Generally speaking, left-brain functions control conscious thought, logical analysis, language and external awareness. Right-brain controls subconscious thought, intuitiveness, and inner awareness. In creative process, both sides of the brain work in unison. Students should be trained to use both parts of the brain.

18

Problem-solving Techniques

Problem-solving need not be confined only to solving numerical problems in mathematics, science, engineering, or technology related courses.

A problem is a situation, condition, or an issue that remains unresolved and makes it difficult to accomplish a desired objective. The situation, condition, or issue is considered as a problem when an individual or group becomes aware of the significant difference between what is desirable and what is actually the situation. Trying to find a solution or an answer to a problem is known as problem-solving.

Problem-solving is a hierarchical process by which a situation is analysed, workable solution arrived at, and corrective action is implemented. The strategic steps involved are:

- Defining or identifying the problem (clarifying the situation, condition or issue) and state the problem clearly in unambiguous terms.
- Analyze causes.
- Generate/identify alternatives/ideas or possible solutions.
- Assess the alternatives in terms of their pros and cons.
- Select/decide the most feasible alternative (solution).
- Identify the steps required to implement the solution.
- Implement the solution.

- Evaluate whether the problem is solved or not.

Problem-solving Techniques: Every problem needs a solution and there are many problem-solving techniques. These are:

- **Affinity Diagrams** (organising ideas into common themes).
- **Ansoff Matrix** (understanding different risks/ implications of different alternatives).
- **Appreciating Objectively** (extracting maximum information from facts and figures, and understanding them).
- **Appreciative Enquiry** (looking at what is going right and solving the problems).
- **Brain Storming** (generating a large number of ideas from a group to solve a problem).
- **Boston Matrix** (focussing on efforts to get maximum returns).
- **Cause and Effect Diagrams (Fish-bone diagram)** (identifying possible causes of problem, and factors that influence the end effect).
- **Core Competency Analysis** (staying ahead and going ahead).
- **Critical Success Factors** (identifying critical factors that really matter for success).
- **Five Whys Technique** (quickly getting to the root of a problem).
- **Flow Charts** (understanding how a process works).
- **Greiner Curve** (surviving the crisis that come with growth).
- **De Bano's Lateral Thinking** (changing/inversing the concepts and perception).
- **Marketing Mix and the Four P's** (understanding how to position oneself in the market).
- **Mckinsey Seven S's** (making sure that all groups of an organisations work in harmony).

- **PEST (Political, Economic, Socio-cultural and Technological) Analysis** (understanding the whole picture/scenario).
- **Porter's Five Forces** (understanding where power lies).
- **Reframing Matrix** (examining problems with distinct viewpoints).
- **Risk Analysis.**
- **Systems Diagrams** (understanding the system, its objectives, components, and how different factors affect one another/outcome).
- **Root Cause Analysis** (identifying various factors, causes of problems, or events).
- **SWOT** (Strengths, Weaknesses, Opportunities and Threats).
- **USP (Unique Selling Propositions) Analysis** (evolving competitive edge).

The SCAMPER Technique

The SCAMPER technique uses a set of directed and idea-spurring questions to suggest alterations/additions to resolve a problem and modifications to something that already exists. The principles of the SCAMPER technique were first formally suggested by Alex Osborn and later structured by Bob Eherle as a Mnemonic tool to increase interest in the perceptive, imaginative, and creative abilities amongst children and adults. The SCAMPER technique also received much attention as a learning tool that fosters awareness drive, fluency, flexibility, and originality. The stimus comes from asking a series of questions that one would normally pose and arriving at answers which will resolve a problem.

SCAMPER stands for:

S – Substitute (components, materials, or people).

C – Combine (mix, combine with other assemblies or services, integrate).

A – Adapt (alter, change the function, use part of another element).

M – Magnify/Modify (increase or reduce in scale, change shape, modify the attributes).

P – Put to other uses/alternative uses.

E – Eliminate (remove the elements, simplify, reduce to core functionality).

R – Rearrange/Reverse (turn inside out or upside down).

Guide to the SCAMPER Technique

Substitute: Think about substituting a part of the product or a stage of the process with something else.

Typical Questions:

- What can be used instead?
- What other materials, ingredients, processes, approaches might be substituted?

Cohesive: Think about combining two or more parts of the product, or stages of the process.

Typical Questions:

- What mix or combination of parts or processes can be combined or integrated?

Adapt: Think about which part of the product or process could be adapted.

Typical Questions:

- What else can be done?
- What ideas can be adopted?

Magnify/Modify: Think about changing the part or modifying the part.

Typical Questions:

- What other size, form, shape, colour can be adopted?
- What can be added?

Put to Other Uses: Think of how you can put the product to another use or how will you reuse them.

Typical Questions:

- What are new ways of using this product/process?

- Which other people will be interested in this product/process?

Eliminate: Think of what will happen if you eliminate this product, part or process.

Typical Question:

- What is that which I can eliminate and what will happen as a result of this?

Re-arrange/Reverse: Think of what will happen if the part/product works in reverse order (or sequenced differently)?

Typical Questions:

- What might be rearranged?
- What other pattern, layout, or sequence I can adopt?
- Can components be interchanged/reversed? If so, what will happen?
- Could roles be reversed?

The SCAMPER technique is used to create original ideas. The creative process thrives on preparation, concentration, illumination, verification, and incubation. The fruitful application depends on the enabling environment of the organization. Engineers/Managers/Supervisors who foster creativity listen and willing to innovate. They must sell their ideas to top management developing persuasive arguments.

19

Creativity and Innovation

Creativity is the process of generating new ideas or concepts, or even devising new ways in which already existing ideas and concepts can be combined. Thus, creativity is a thought process, a fertile thought process.

Innovation, on the other hand, is the conversion of these creative ideas and concepts into commercially viable products or processes that meet specific organizational goals.

Creativity and innovation are not separate entities, they in fact go together—creativity without innovation renders the so called idea useless in the absence of its transformation to new product or process.

In the present context of today's events, creativity and innovation should be viewed as a continuous improvement process of devising new ways and products. Innovation and research would play a key role in the future.

Characteristics of innovative leaders:

- Fast and action oriented.
- Deeply immersed and involved in progressive changes.
- Future-obsessed.
- They are highly motivating and inspiring.
- Passionate about their dreams and goals.
- They don't take "No" for an answer.

- They are Institution and Organization builders.
- High-gain and risk-takers; they don't think small. Their plans are always grandiose; they think big and their actions are daring.

Innovative leadership is the highest form of leadership because of the positive impact and huge contributions they have made in the lives of their engineers, employees, customers, and community. Innovative leaders take their companies and organizations to far greater heights and to exciting destinations.

We need innovative leaders who can transform the people they lead, transform the people they interact within their organizations. The people, then, go on to transform the business landscape and, thus, the entire organization.

Innovative leadership calls for challenging current processes, envisioning a change, developing and inspiring a shared vision, and modeling and leading the way forward. These leaders have charisma, have vision, propagate intellectual stimulation, and above all inspire people across all disciplines.

An innovative leader achieves excellence in a different way than he once envisioned. Instead of working his way through an existing ladder, he is building a new one.

People Skills

An independent study by Stanford Research Institute and Carnegie Mellon University, U.S.A. on CEOs, who are indisputable leaders of their organizations, revealed that long-term success depends 75% on people skills and 25% on technical skills. Another research by Harvard University, U.S.A. found that organizations lose 10-15% of their business due to the lack of soft skills in their employees. Innovative leaders need to take cognizance of this fact.

Who is a "A Team Player"?

It is very difficult to describe a team player. We look for intellect and personality, but we also look for internal qualities like motivation, creativity, and drive. We want people with a desire to win, who can leverage their own talents as part of a team, not someone who is focused on pursuing individual achievement. It is difficult to list the right combination of adjectives. You can see it in athletes. The best football players are not necessarily the fastest.

Product Innovation and Product Leadership

Product innovation and product leadership are two important ingredients for a company to become successful. Michael Tracy in his book, "Discipline of Market Leaders", talks about three different strategies—product leadership, cost leadership (operational efficiency), and customer intimacy. Continuous innovation is necessary in the global economy. Many successful companies have embraced product leadership from day one, with a high degree of customer intimacy.

Developing a culture of Creativity and Innovation involves:

- Brainstorming for breakthrough ideas.
- Quickly try to convert your ideas into new gadgets, devices, products, and services.
- Anticipate what your customers or buyers want.
- Think differently; think "out-of-the box", think in terms of simple operation, better quality, faster delivery, and above all at reduced cost.
- Working with cross-disciplinary experts and teams. Ask the opinion of experts from other areas.
- Substituting with new materials and/or new design.
- Modifying existing design with a better design.
- Combining two or more existing parts and replacing them with a new part (i.e., a composite part).
- Adjusting, reversing/inversing, and eliminating the part altogether.
- Imagine without any constraint and develop/create a new configuration.
- Ask yourself is there an analogy exists in the nature or natural process?
- Question the status-quo; What next? Why not? What else? How else, What if?
- Read the book on Lateral thinking by Edward De Bono.

Innovation

- "If you can dream it, you can do it"—Walt Disney.
- Innovation is not merely a new or novel concept.
- Innovation is a sustainable idea, is a process, and is a manifestation of originality—originality in thinking and originality in conceptualising natural inventiveness of the mind.
- Stretching the mind little further and farther than the normal thresholds has resulted in great innovations.

Much of the world-changing innovations that we are witnessing in the 21st century emerges at and from the intersection of many disciplines:

> Science, Art, Business, Design, Technology, Education, Medicine, and beyond.

This innovative thinking occurs most frequently when you have the opportunity to step away from your regular work and learn about, discuss and explore how to integrate diverse and seemingly unrelated ideas, linkages, and connections. If you want to become an innovator you should experience the world of limitless creativity that will inspire and energise you.

History of Innovations

> GPS, Bar Code Scanners, CAD, Improved Bio-fuels, MRI Technology, Fetal Monitoring, Transistors, Forensic DNA Analysis, Google Search Algorithm, NSFNET-Pre-Internet, Improved Weather Forecasting.

Global Challenges

> Climate change, Global health, Food security, Water security, Bio-diversity, Population security, Energy security.

What Does It Mean to Innovate

1. Encourage unproven or untested ideas. Encourage creative ideas amongst the employees.
2. Listen to diverse points of view. Reach out to employees for ideas on improving efficiency.
3. Create meaningful work environment. Motivated employees help to improve the productivity of the organization.

4. Energise and encourage the talented employees.
5. Encourage flexibility and employees involvement, motivate the employees to do their best.
6. Develop talents amongst the employees by organising training programmes.
7. Offer sound performance incentives and encourage collaboration and team work.
8. Respect diversity of opinions amongst the employees. To solve problems, consider points of view from multiple perspectives.
9. Important strategic and operational decisions are quickly translated into actions.

In a rapidly changing environment, innovation is the key. Best-in-class companies nurture talent and foster innovation in their employees.

Innovation Strategies

1. Think out-of-box.
2. Think laterally (learn to think differently).
3. Develop and cultivate creativity.
4. Acquire new skills and talents.
5. Look for new patterns and value creations.
6. Consider "Problems" as "Challenges".
7. Convert the "Challenges" into "Opportunities".
8. Dare to be different.
9. Challenge the "Status quo", no matter how perfect the "Status quo" is.
10. Have "Fire in the Belly" to convert ideas into reality.
11. Develop a passion for excellence.
12. Unshackle the human spirit of creativity.
13. Minimise the innovation cycle time.
14. Maximise the innovation success rates.
15. Be competitive.
16. Convert the "Constraints" into "Advantages".
17. Learn from failures.
18. Foster high performance work culture.

20

How Innovations have Evolved?

The world is replete with innovations that have now become an essential of our daily lives. Some of the breakthrough innovations of the past two centuries are the Computer, Microphone, Telephone, Electric Bulb, Radio, Airplane, Television, Rocket, Barcode, Laser, Mobile Phone, and the World Wide Web. The various innovations described herein illustrate how these innovations have evolved over the years and changed the way we live.

Engineering Innovations:

1. **The Computer:** Changed the way world WORKS.
2. **The Microphone:** Changed the way world SPEAKS.
3. **The Telephone:** Changed the way world COMMUNICATES.
4. **The Electric Bulb:** Changed the way world SEES.
5. **The Radio:** Changed the way world BROADCASTS.
6. **The Airplane:** Changed the way world FLIES.
7. **The Television:** Changed the way world is ENTERTAINED.
8. **The Rocket:** Changed the way world EXPLORES.
9. **The Barcode:** Changed the way world IDENTIFIES.
10. **The Laser:** Changed the way world uses LIGHT.
11. **The Mobile Phone/(The Cell Phone):** Changed the way world COMMUNICATES on the move.

12. **The World Wide Web:** Changed the way world CONNECTS.

1. **The Computer:** Changed the way world WORKS.

Charles Babbage originated the concept of a programmable computer in 1822. Since then the computer has evolved from a simple mechanical device, which was used for ordinary calculations to a sophisticated electronic system which can perform millions of calculations per second. Today, we cannot imagine life without computers.

Now, computers have evolved from standalone systems to embedded computing devices that can be found in a wide range of applications from toys and MP3 players to fighter aircrafts and industrial robots.

2. **The Microphone:** Changed the way world SPEAKS.

The first Microphone was invented by Emile Berliner in 1876 and was initially used as a telephone voice transmitter.

Over the years, the Microphone has evolved and has found use in various applications ranging from tape recorders to televisions broadcasting and in computers for recording voice, speech recognition, and VOIP, (voice operated internet protocol).

3. **The Telephone:** Changed the way world COMMUNICATES.

After the invention of the telegraph, the next major challenge was to transmit the human voice across large distance.

Alexander Graham Bell accomplished this by inventing the telephone in 1876. Over the past few decades, digital telephony has improved the capacity and quality of the telephone network.

Today, the telephone has become one of the most common appliances in the world with over 4 billion mobile and fixed-line subscribers.

4. **The Electric Bulb:** Changed the way world SEES.

Thomas Alva Edison, credited for this life altering invention in 1879, attended the school for just three months.

It was Edison's mother who taught him at the home itself and played the important role of a teacher in his life.

Edison dedicated his inventions to his mother and recalled: "My mother was the making of me. She was so true, so sure of me; and I felt I had something to live for, someone I must not disappoint."

5. The Radio: Changed the way world BROADCASTS.

The Radio, which was invented by Guglielmo Marconi in 1896, replaced the wired telegraph communication system.

The first major triumph for this invention was the transatlantic wireless signal transfer from England to Newfoundland (Canada) in 1901.

In its early years, Radio was used to send messages and communication between armies during wars.

Today, the Radio facilitates the communication of News, Education, and Entertainment to millions of people across the world.

6. The Airplane: Changed the way world FLIES.

This extraordinary feat was accomplished by two young bicycle mechanics from the USA, Wilbur Wright and Orville Wright, who invented the first airplane, which took off on 17th December, 1903. The first flight lasted all of 12 seconds and covered a distance of 37 meters. Today, this invention has shrunk the world and over 3 billion passengers travel by air every year.

7. The Television: Changed the way world is ENTERTAINED.

John Logie Baird first encountered the word "Television", which was the latest term, a concept, published in a magazine in 1924. This ignited in him the thought of actually inventing the device.

On 1st July, 1941 the first ever commercial broadcast took place on Television in America.

The invention, which began as a semi-analogue, mechanical black and white broadcast system has undergone tremendous developments to the sophisticated colour, electronic Television of today.

8. The Rocket: Changed the way world EXPLORES.

For centuries rockets were principally used as weapons, for signaling, and for fireworks displays.

Robert Hutchings Goddard is credited with creating and building the world's first liquid-fuelled rocket in 1926.

It was only in the 20th century it became the enabling technology of the space age, including setting foot on the moon and the hope to create a space tourism industry with manned spaceflights.

9. The Barcode: Changed the way world IDENTIFIES.

Norman Joseph Woodland and Bernard Silver redefined the concept of unique identification with the invention of the Barcode in the year 1952.

Today, Barcodes are everywhere, used in various industries such as book sales, retail, airlines, car rentals, entertainment, and automobiles, to name just a few.

10. The Laser: Changed the way world uses LIGHT.

Lasers have found use in almost every field from medical and military applications to commercial and cosmetic uses.

Theodore Maiman, who invented the first working laser was able to develop his ideas on the concept of microwave-optical measurements and built the first working Laser, which he finally accomplished in 1960.

11. The Mobile Phone: Changed the way world COMMUNICATES on the Move.

Martin Cooper invented the first Mobile Phone that weighed 2 kg with a battery that lasted for 10 minutes in 1973, since then, Mobile Phones have become a lot smaller over the years but they're bigger than ever in functionality and popularity.

Mobile devices have now evolved to powerful pocket computers, game machines, and multimedia-playback devices.

12. The World Wide Web: Changed the way world CONNECTS.

Tim Burner-Lee published a paper in 1989 that detailed how the World Wide Web would operate. He created a hyperlinked database system for use in connecting databases across a closed network. Ever since, the World Wide Web has been exploding into all aspects of our work, studies, leisure, and life. Today, the internet, which the World Wide Web rides on, has over 1.5 billion users and more than 100 billion live Web pages.

21

Leadership

There are many ways of defining Leadership.

- Basically, it is about getting things done through people.
- Leader shows the way and ensures that everyone gets there.
- Leadership is about encouraging and inspiring individuals and teams to give their best to achieve a desired result.
- Managers as leaders have to achieve the task. That is why they and their teams exist.
- Leader's role is to ensure that the team's purpose is fulfilled. If it is not, the result is frustration, disharmony, criticisms, and eventually, perhaps, disintegration of the group.
- Achieving the task involves leaders in getting answers to the following questions:
  - What needs to be done and why?
  - What results have to be achieved?
  - What problems have to be overcome?
  - Is the solution to these problems straightforward or is there some ambiguity?
  - Is this crisis situation?
  - What is the time frame for completing the task?

In the light of this analysis leaders have to take whatever steps are necessary to motivate the individuals in their team and also, importantly, to develop good team work.

Leadership is the ability to effectively combine individuals and resources together to accomplish things that would be virtually impossible to achieve alone.

Personal leadership is the ability to visualise a goal, to embrace the values of that goal, and maintain a positive perspective in a self-disciplined environment until the goal is achieved.

Organizational leadership is the ability to articulate a vision, to embrace the values of that vision, and nurture a positive environment where every one can reach the organization's goals and their own personal needs.

Attributes of Good Leadership

- Self-control.
- Keen sense of fairness and justice.
- Good decision-making.
- Good planner (plan his work and work his plan).
- Willingness to do more than what is just required.
- Pleasing personality.
- Masters details.
- Sympathetic and has good understanding.
- Willingness to assume full responsibility (not merely for success, but also for failure).
- Cooperate.
- Exhibits confidence and courage.
- Gets the job done.
- Inspires others to join him in working towards a common goal.
- Willingness to take risk.
- Compassionate and has deep concern for the welfare of others.

- Good leader's actions are self-generated and not merely reflective of the actions of others.

Indispensable Qualities of a Leader

(*Courtesy*: John Maxwell, Leadership Outline)

1. Character (good qualities and features).
2. Charisma (powerful personal quality that attracts other people).
3. Commitment (promise to do).
4. Communication (process of expressing ideas and feelings).
5. Competence (ability to do well).
6. Courage (bravery/opposing without showing fear).
7. Discernment (ability to show good judgement).
8. Focus (give attention to one particular thing).
9. Generosity (willing to give more than what is necessary).
10. Initiative (ability to decide and act on your own).
11. Listening (paying attention to what you hear).
12. Passion (strong feeling of love/enthusiasm).
13. Positive attitude (positive way of thinking and feeling).
14. Problem-solving (finding a solution/finding ways of dealing with problems).
15. Relationships (ways in which people behave towards each other).
16. Responsibility (taking care of).
17. Security (activities involved in protecting).
18. Self-discipline (ability to make yourself do).
19. Servant-hood (ability to render service with devotion).
20. Teachability (able to learn by being taught).
21. Vision (ability to think about/plan the future with great imagination and intelligence-foresight).

Important Characteristics of Good Leaders

(*Courtesy*: "Quality is Free" by Philip B. Crosby)

These are:

- Importance of listening,
- Cooperating and team work,
- Helping others,
- Communicating,
- Creating (be creative),
- Implementing a solution,
- Learning (be a life-long learner),
- Leading others, and
- Following the leader.

There are other important characteristics like integrity and compassion. These are vital and fundamental facts of one's character and they can be learned or vastly improved. You cannot succeed in the company or business without them. As a matter of fact it is difficult to think of any profession you would want to be part of that wouldn't demand a great deal of integrity and compassion.

If you want to accomplish more than what you have accomplished in this complex world you would have to be more than the next person. Take advantage of the assets you have.

Importance of Listening

You can pay people no greater honour than to actually absorb the content and intent of what they have to say. Not just the message as you have heard superficially; but you must concentrate on what is being said behind the words. Listen really to the speaker. Is the person in trouble? Does he/she want help? Is he/she saying one thing and meaning another? What is really going on?

You must listen and question until you are sure you understand the message. You will discover an interesting world. Once people find that you are willing to take the time to understand them, they will take the time to arrange thoughts into concepts that are easier for you to comprehend. You may even create some listeners of your own. Then you can reach a state where true communication takes place. You are actually transmitting and receiving real meanings.

Once you have a basis for eliminating or preventing misunderstandings, success is at hand. Nothing is more important than true understanding.

Cooperating and Team Work

Being part of a team is not a natural instinct or human function; it is learned. Participants in a contact sport, such as football or hockey learn quickly that the rest of the team evaluates personal worth by evaluating a team member's determination to cooperate. The player must not only makes the play to be executed and successful, but protect others as well in the process.

To cooperate with others in accomplishing a task is one of our nobler achievements. It is also one of the most productive aspects of the manager with a style. The whole reality is greater than the sum of its parts. Synergism really does work, if you take care of it.

Cooperation doesn't mean that you have to abandon or sacrifice any of your personal traits or standards. You can leave one team and join another. Make certain that the team is worthy of you, just as you must be worthy of the team. Expect only the best. Don't settle for less. A sure way to successful cooperation is to help build a winning team from one that is not doing well. Bring people together, manage, and lead. The same principles apply in relationships with individuals. Those who are known to work with others for a common goal are known as trustworthy people.

But most of all, do not be uncooperative. If that is so, you will be ignored. Ignored managers cannot cooperate with any one.

Helping Others

Helping and cooperating are not the same. Helping involves giving, without direct expectation of immediate return. It is letting some one lean on you without getting to lean back. In cooperating you both are together, leaning on something for mutual support.

In corporate life, helping has to do with smoothening the road for others, whether they know you are doing it or not. It is being available when someone needs help. It is putting in a good word for someone who deserves it. It is like offering a discrete personal loan or gift without making big deal out of it.

To help in a positive manner, you must be genuinely interested in people and results. True help is truly unselfish. You must have the patience to wait until the individual recognises that he/she really does need assistance. Help must be carefully thought out. If you are prepared to offer assistance properly, you can guide him and counsel him as well. If you are confident that your efforts to help are based on a genuine concern for the individual, and are not to fulfill your own ends, then the help will be accepted.

Communicating

Techniques such as public speaking, personality improvement courses/sessions, transactional analysis, appropriate and effective use of body language—these and many other activities have been undertaken by people in the hope of improving the way they communicate.

Usually they achieve a better understanding of themselves, which is what most of these activities are all about, but they fail to improve their communication. They don't understand or believe a fundamental truth that you have to have something to say.

Communicating is a characteristic functional activity because there is much more to conversing your thoughts than just speaking. Your dress/clothes, your physical manner, your personality, your cleanliness, and many other things transmit a lot about you, often so loudly and cleanly that the words, you choose to say may not be heard, at all. How are you communicating should not be left to chance.

There are three basic communication actions: writing, prepared speaking, and conversation.

(a) **Writing** is putting one's thought clearly, short, and direct (straight); long and drawn out message is undesirable. Writing helps you bring out your thoughts together coherently so that others can understand them. You may be struggling continuously to make yourself understood. Writing is the hardest in the art of communication. When you speak to an audience, you can get a little feedback as you go along. You can detect confusion, boredom, or drowsiness. Reading is a private act, and you can never know if someone understands what you have written unless that person shows evidence of having understood (or misunderstood). Being understood is a hard work. But, if you are successful at it, you can achieve more than you ever dreamed.

(b) **Prepared speaking** is a rewarding communication method; and it takes a lot of efforts. But a really good speech doesn't seem well prepared, and yet it takes so long to produce one. Preparing formally for any speech is always desirable. Training programmes on effective public speaking are regularly organised for faculty/ employees in the organisation/institution. In addition, professional help is advised for learning how to construct a good speech; and this gives confidence.

(c) **Conversation**: Many people feel that conversation requires no preparation and attention. Smart executives, engineers, and managers listen to themselves and learn how to improve. They learn things like how important it is to look into the eyes of the other person; like when to touch them and when not; like when not to be clever or funny; like when just to remain silent.

All these communication factors, such as writing, speaking and conversing—project a picture of you on everyone's mental wall; what they see is what they think you are. You may feel that you are something else, and may actually not be what is being projected to others.

Creating (Be Creative)

You should learn to be creative. You can create solutions to complicated problems by being the only one to break that complicated problem down to its basic elements/causes. Once these causes are exposed, a creative solution appears—if you can recognise it. Unfortunately, many people give up before that New and original solutions are the result of hard work.

Problem-solving is an acquirable skill. Many techniques and concepts for learning creativity are available. The most practical technique, however, is to go and ask the people who are involved with the problem on the job/shop floor; not the management people (like engineers/managers), but the working people, i.e., people on the job. The operators are delighted that someone cares about their opinion. They share with you to create a solution. At the same time, make certain that they also get their due credit.

One other source of creativity is to expand on the ideas of others, or to work on the old ideas and update them. Give credit where it is due.

Don't tell that you lack creative activity. Everyone is endowed with creativity which can be nurtured and developed.

Implementing a Solution

The most valuable manager is one who can first create and then implement. You can implement others ideas/concept also or add your own improvements to already existing methods. Most managers can only follow directions; but real implementers are appreciated and respected by others.

You have to get the job done: Understanding the basic problems, determining a practical solution, convincing others about the new solution, and then getting it done is the way to move forward. Many quality/productivity improvements are like that. Don't expect commitment that every one will cooperate. If you can get it, it is wonderful.

Learning (Be a Life-long Learner)

Learn your subject. The more information you have about everything, especially people, technology, and finance, the more you can accomplish. When you have an answer for everything, you have stopped learning. Hence, it is essential that you should not stop learning. It is important that you read new information and listen. Read at least a few magazines/books every week or month. Don't follow the same routine. Keep yourself alive.

It is not always possible to know what you need to learn. Therefore, you have to constantly seek out new experiences and exposures. You should know the latest development; know how to deal with them or the effect they create. Relate the new developments to those of the past. New activities are stimulating and challenging; they exercise the learning muscles in you continuously. It will create positive feelings, thoughts, and enthusiasm in you.

Leading Others

Leaders communicate by giving clear and unambiguous direction and by setting examples as evident. Leading means stating the objectives in a clear way that is precisely understood, ensuring commitment of individuals to those objectives, defining how the results will measured, and then by providing impetus to get things done. Leading is a hard and unending job. To lead, you have to keep everyone busy. The leader should never run out of ideas/jobs, for the people who are working under him and seeking his guidance.

The conventional thoughts about the problems of being a leader: the long hours, the problems of choosing between those who are coming up and those who are going down. Leader has to survive such problems.

Leaders' great mistake is the belief that only they can set the goal, their belief that they know how to do anything.

Open-mindedness—an approach in which ideas are floated, widely discussed, and guidelines are proposed openly—is the only cure. A closed leadership doesn't grow. Solo shows are also not conducive for good leadership.

Following the Leader

It is not easy to be a good follower. A good follower wants the same results that the leader wants. The follower doesn't necessarily need to have identical goals, but the desire for the same results must exist.

A good follower takes time to make certain that he or she understands the intent (intention) of the leader's request as well as his words. Good followers, who want to expand their sphere of influence, get promoted and move into higher activities. They expand into an area where there is a lot of scope to act and do it with the full understanding of the leader. They do not compete with the leader. Be a good follower, leading will take care of itself.

Quotes on Leadership

- "A genuine leader is not a searcher for consensus, but a moulder of consensus."—Martin Luther King
- "Management is doing things right,
  Leadership is doing the right things."—Peter Drucker
- "Innovation distinguishes between a leader and a follower."—Steve Jobs
- "Setting an example is not the main means of influencing others;
  It is the only means."—Albert Einstein
- "I suppose leadership at one time meant muscles; but today it means getting along with people."—Mohandas K. Gandhi
- "To be leader, you have to lead human beings with affection."—J.R.D. Tata
- "A good objective of the leadership is
  To help those who are doing poorly to do well, and
  To help those who are doing well to do even better."
  —Jim Rohn
- "Leadership is the art of getting someone else to do something you want done because he wants to do it."
  —Dwight D. Eisenhower
- "I have a different vision of Leadership.
  A Leadership is someone who brings people together."
  —George W. Bush

- "Management is the efficiency in climbing the ladder of success;
  Leadership determines whether the ladder is leaning against the right wall."—Stephen Covey
- "It is better to lead from behind and put others in front, especially when you celebrate victory when nice things occur."—Nelson Mandela
- "You take the front line where there is danger. Then people will appreciate your leadership."—Nelson Mandela
- "Leadership and learning are indispensable to each other."—John F. Kennedy
- "A good leader leads his followers from where they have ever been to where they have never been."—Henry Kissinger.

22

Our Leaders and Intellectuals

Mohandas K. Gandhi

Gandhi was referred to as Mahatma or "Great Soul," for the first time by Rabindranath Tagore.

Gandhiji was the pre-eminent political and ideological leader of India during the Indian Independence Movement. A pioneer of Satyagraha, or resistance to tyranny through mass civil disobedience—a philosophy firmly founded upon Ahimsa, or total non-violence. Gandhi led India to independence and inspired the civil rights and freedom movements across the world. In India, he is officially honored as the "Father of the Nation".

As a leader, he stressed the importance of continuous development of individuals to propel them to achieve greater heights. He possessed the quality of not coercing people by convincing them to follow him. Till his last breath, he strived to create a nation where all people irrespective of their religion, caste, creed, or gender live together in peace and harmony.

Rabindranath Tagore

Rabindranath Tagore was one of the greatest intellectuals of modern India.

He has authored many poems, novels, stories, songs, dance-dramas, and essays. He wrote "Jana Gana Mana", which became India's National Anthem.

He was awarded the Nobel Prize for Literature in 1913 and was the first Indian to receive this Honour.

He participated in the Indian nationalist/freedom movement in a rational and visionary way. He promoted spiritual values and the creation of a new world order founded on multiculturalism, diversity, and tolerance.

Dr. S. Radhakrishnan

Sarvepalli Radhakrishnan was a great philosopher and statesman. He was the first Vice President of India (1952-1962) and subsequently the second President of India (1962-1967).

He was born in an humble Telugu family in a small town in Tamil Nadu.

He was a bright student throughout his academic life.

Dr. Radhakrishnan served in many educational institutions in various capacities. When he became President, some of his students and friends requested him to allow them to celebrate his birthday, which fell on September 5th. He expressed the view that instead of the celebrations, he would consider it a privilege if it was observed as "Teachers' Day". Since then September 5th is being celebrated as "Teachers' Day" throughout in India.

He was one of India's most influential scholars of comparative religion and philosophy, and built bridge between the East and the West by showing that the philosophical system of each tradition are comprehensive within the framework of each other.

Mother Teresa

Perhaps no modern figure has come to symbolise compassion more than the small, elderly Albanian nun who later became an Indian Citizen and was known to millions as Mother Teresa. Early in her life, in 1950, she received approval from the Vatican and started her own religious order, "Missionaries of Charity" whose vision was to love and care for the uncared.

She gave herself completely to her cause, made innumerable sacrifices, and was deeply committed to bring about change. She was someone who was not afraid to do the right thing or motivated others to do so. Through her action and dedication, she motivated many people to contribute to helping the poor of the world. Mother Teresa was awarded the Nobel Prize for Peace in 1979 and the Bharat Ratna, India's highest civilian honour, in 1980. She has left a deep imprint on humanity.

J.R.D. Tata

One of the pioneering industrialists of India Jehangir Ratanji Dadabhoy (J.R.D.) Tata became the Chairman of the Tata Group in 1938. Under him, the Tata Group diversified into a large number of corporates including airlines, automobiles, hotels, locomotives, chemical, pharmaceuticals, financial, and consultancy services. As a leader, he believed that he had to motivate individuals and inspire respect.

He was also a pioneering aviator and brought commercial aviation to India. The Tata's provided philanthropic support to India's painters, sculptors, and performing artistes.

Galileo Galilei

Galileo Galilei was an Italian physicist, mathematician, astronomer, and philosopher who played a major role in the scientific revolution. Galileo is called "the Father of Modern Science".

Galileo made original contributions to the science of motion through an innovative combination of experiments and mathematics. Galileo was one of the first modern thinkers to clearly state that the laws of nature are mathematical. His theoretical and experimental work on the motions of bodies was a precursor of the classical mechanics developed by Sir Isaac Newton. He constructed a telescope with which he studied lunar craters and discovered four moons revolving around Jupiter.

His conviction that the Earth and the other planets revolved around the sun was controversial within his lifetime. He was tried, found "vehemently suspect of wrong idea", forced to withdraw his theory, and spend the rest of his life under house arrest. It was while Galileo was under house arrest that he wrote one of his finest works, *Two New Sciences*. Here he summarised the work he had on the two sciences, now called Kinematics and Strength of Materials.

Abraham Lincoln

Abraham Lincoln, the 16th President of the United States of America, was the son of a Kentucky soldier and had to struggle for living and for learning. Lincoln made extraordinary efforts to attain knowledge while working on a farm, splitting rails for fences, and working in a store at New Salem, Illinois, U.S.A.

He served as President of United States of America from March 1961 until his assassination in April 1965. He successfully led his country through a great constitutional, military crisis—the American Civil War preserving the American Union and ending slavery.

Lincoln never let the world forget that the Civil War involved an even larger issue. This he stated most movingly in dedicating the military cemetery at Gettysburg: "that we here highly resolve that these dead shall not have died in vain-that this nation, under God, shall have a new birth of freedom–and that government of the people, by the people, for the people, shall not perish from the earth".

Helen Keller

Helen Keller was a truly extraordinary woman. Keller was less than two years old when she was affected by a deadly fever, which left her blind and deaf. When she was seven years old Anne Sullivan herself visually impaired and only 20 years old, arrived at Keller house in Alabama, USA as her tutor and this marked the beginning of a relationship for nearly five decades.

In 1904, at the age of 24, Keller graduated from Radcliffe, becoming the first deaf blind person to earn a Bachelor of Arts Degree. Keller later became a world-famous speaker and author. She is remembered as an advocate for people with disabilities and numerous other causes. She fought for voting rights of women and abolition of war, and was a radical socialist and a birth control supporter. She co-founded the Helen Keller International (HKI) organization and also helped to found the American Civil Liberties Union (ACLU). Helen Keller's triumphs over the adversity of being a blind person have led to her becoming a role model for millions of disadvantaged people and an inspiration for all humanity.

Albert Einstein

Albert Einstein revolutionised Physics with his special and general theories of relativity. Einstein always appeared to have a clear view of the problems of physics and had the determination to solve them. He had a strategy of his own and was able to visualise the main stages on the way on his goal. He regarded his major achievements as mere stepping-stones for the next advance.

He was honored with the Nobel Prize for Physics in the year 1921 for his discovery of the law of the photoelectric effect. He used his leadership on scientific issues to gain a hearing for his views on politics at the highest level of the American Government. Though he played a major role in initiating the US atomic bomb project during World War II, this was a defensive measure and he was against using nuclear capability as weapon. Later, together with Bertrand Russell, Einstein signed the Russell-Einstein Manifesto, which highlighted the danger of nuclear weapons.

Martin Luther King

Martin Luther King was one of the greatest leaders of the African American Civil Rights Movement. His leadership was enhanced by his powerful oratory skills. His passionate and moving speech viz. "I have a Dream" which he delivered at a Civil Rights Washington Rally galvanised people from all walks of life, races, and paved the way for anti-racist legislation in the USA. He provided new leadership for the Civil Rights Movement by combining the ideals of Christianity and the non-violent preaching of Mahatma Gandhi.

In 1964, Martin Luther King became the youngest person to receive the Nobel Prize for his work to end racial inequality through civil disobedience movement and other non-violent methods.

He was just thirty-nine years old when he was assassinated on 4th April, 1968 in Memphis, Tennessee, USA.

Peter Drucker

Peter F. Drucker was a professor, author, management consultant, and "social ecologist", who explored the way human beings organise themselves and interact, much the way an ecologist would observe and analyse the biological world. Besides the strategic aspect of business, Drucker also studied about how public and private organizations operate morally and ethically within a civil society.

Drucker authored over 35 books and a large number of articles on management, executive leadership, economics, politics, and society. He has been the most popular management philosopher of the last century and was a consultant to General Motors, Intel, and General Electric's, USA.

Drucker believed that an executive who establishes the spirit of performance in daily life is a leader in his or her organization.

Steve Jobs

Steve Paul Jobs is undeniably one of the greatest innovators of all time. He has left his mark on no less than four industries: personal computers with Apple II and Macintosh, music with iPod and iTunes, phones with iPhone, and animation with Pixar. He has changed millions of lives by making technology easy-to-use, exciting, and beautiful.

The middle-class bohemian kid with no college education built a computer empire and became a multi-millionaire in few years. He was then fired from his own company before coming back a decade later to save it and turn it into one of the world's most influential corporations with millions of fans around the world. He has been justifiably hailed as visionary leader whose dream of merging the worlds of technology and creative design has made Apple one of the most innovative companies of the world.

23

Understanding Life

The Twelve Great Riches of Life

Grade yourself: Perfect, Fair, or Poor.

1. A positive mental attitude (i.e., state of mind).
2. Sound physical health (i.e., health consciousness).
3. Harmony in human relations (i.e., harmony with oneself within and harmony with others).
4. Freedom from fear (refer: Seven Basic Fears).
5. The hope of future achievement.
6. The capacity for faith (spiritual quality).
7. Willingness to share one's blessings.
8. A labour of love (i.e., joy of self-expression).
9. An open mind on all subjects.
10. Self-discipline.
11. The capacity to understand people (inspired by nine basic motives of life).
12. Economic security (i.e., the tangible portion of the twelve great riches of life).

Embrace the twelve great riches of life, apply them in your daily life, and you will become a success—for success is nothing more than the attainment of these twelve blessings.

The Nine Basic Motives of Life

1. The emotion of LOVE.
2. The emotion of SEX.
3. The desire for MATERIAL GAIN.
4. The desire for SELF-PRESERVATION.
5. The desire for FREEDOM OF BODY AND MIND.
6. The desire for RECOGNITION and SELF-EXPRESSION.
7. The desire for PERPETUATION OF LIFE AFTER DEATH.
8. The emotion of ANGER.
9. The emotion of FEAR (refer: Seven Basic Fears).

The Seven Basic Fears

1. The fear of POVERTY.
2. The fear of CRITICISM.
3. The fear of ILL HEALTH and PHYSICAL PAIN.
4. The Fear of LOSS OF LOVE.
5. The Fear of LOSS OF LIBERTY-FREEDOM.
6. The Fear of OLD AGE.
7. The Fear of DEATH.

The Eight Guiding Princes

1. Prince of Financial Prosperity.
2. Prince of Sound Physical Health.
3. Prince of Peace of Mind.
4. Prince of Hope, and } operating as twins
5. Prince of Faith. }
6. Prince of Love, and } operating as twins
7. Prince of Romance. }
8. Prince of Overall Wisdom.

How to Become a Friendlier Person?

- Begin in a friendly way.
- Keep a smiling face.
- Remember that a person's name is the sweetest to him and most important sound in any language.
- Be a good listener.
- Encourage others to talk about themselves.
- Don't criticise, condemn, or complain.
- Give an honest and sincere appreciation.
- Become genuinely interested in other people.
- Talk in terms of other person's interests.
- Make the other person feel important—and do it sincerely.
- The only way to get the best of an argument is to avoid it.
- Show respect for the other person's opinion. Never say, "You're wrong".
- If you are wrong, admit it quickly and emphatically.
- Let the other person do a great deal of the talking.
- Let the other person feel that the idea is his or hers.
- Try honestly to see things from the other person's point of view.
- Be sympathetic with the other person's ideas and desires.
- Dramatise your ideas.
- Begin with praise and honest appreciation.
- Call attention to people's mistakes indirectly.
- Talk about your own mistakes before criticising the other person.
- Give an opportunity to the other person to save face.
- Praise the slightest improvement and praise every improvement.

- Be hearty in your approval and agreement, and lavish in your praise.
- Encourage the other person. Make the fault seem easy to rectify.
- Make the other person happy about doing the thing you suggest.

Fifty-Four Major Causes of Failure

1. The habit of drifting with circumstances, without definite aims or purposes.
2. Unfavourable physical heredity at birth.
3. Meddle-some curiosity in connection with other people's affairs.
4. Lack of a definite major purpose as a life goal.
5. Inadequate schooling.
6. Lack of self-discipline, which generally manifests itself through excesses in eating, drinking, and sex indulgence, and indifference towards opportunities for self-advancement.
7. Lack of ambition to aim above mediocrity.
8. Ill health, generally due to wrong thinking, improper diet, and lack of physical exercises. (Keep in mind, however, that some people, such as Helen Keller, have made themselves of great service to others, in spite of their own incurable ailments.)
9. Unfavourable environmental influences during childhood. It has been researched that major fundamentals of character have been well formed in the individual by the time he is seven years of age.
10. Lack of persistence in carrying through to a finish that one starts.
11. A negative mental attitude as a fixation of habit.
12. Lack of control over the emotions of the heart.
13. Desire for something for nothing, usually expressed in the habit of gambling.

14. Failure to reach decisions promptly and definitely, and to stand by them after they have been made.
15. One or more of the seven basic fears.
16. Wrong selection of a mate in marriage.
17. Over-caution in business and professional relationships.
18. Lack of all forms of caution.
19. Wrong choice of associates in business or professional pursuits.
20. Wrong selection of a vocation, or total neglect to make a choice.
21. Lack of concentration of effort on the task at hand at a given time.
22. Habit of indiscriminate spending, without a budget control over income and expenditures.
23. Failure of budget and use time to best advantage.
24. Lack of controlled enthusiasm.
25. Intolerance—a closed mind based particularly on ignorance or prejudice in connection with religious, political, and economic subjects.
26. Failure to cooperate with others in a spirit of harmony.
27. The possession of power or wealth not based on merit or not earned.
28. Lack of the spirit of loyalty to those to whom loyalty is due.
29. Egotism and vanity not under control.
30. Habit of forming opinions and building plans without basing them on first-hand knowledge of the necessary facts.
31. Lack of vision and imagination sufficient to recognise favourable opportunities.
32. Unwillingness to go the extra mile in rendering service.
33. The desire for revenge for real or imaginary injuries by others.

34. The habit of conversing in terms of vulgarity or profanity (showing lack of respect).
35. Unsocial attitude towards one's constituted (constitutional) authorities of government.
36. The habit of indulging in negative gossip about the affairs of other people.
37. Unbelief in the existence of Infinite Intelligence (God!).
38. Lack of knowledge of how to engage in prayer so as to bring positive results.
39. Failure to benefit by the counsel of others whose experience one often needs.
40. Carelessness in the payment of personal debts.
41. The habit of lying or unduly modifying the truth.
42. The habit of offering criticism where it has not been invited.
43. Over-extension in connection with the incurring of indebtedness.
44. Greed for material possessions one does not need.
45. Lack of self-confidence of adequate proportion for the fulfillment of one's chosen objectives.
46. Alcoholism or drugs.
47. Over-indulgence in smoking, especially the chain-smoking cigarette habit.
48. The habit (of laymen) of serving as their own lawyers in connection with contracts and legal matters.
49. The habit of endorsing other people's idea when the risk in not justified.
50. The habit of procrastination—putting off until tomorrow that which should have been attended to day before yesterday.
51. The habit of running away from unpleasant circumstances instead of mastering them.
52. The habit of talking too much and listening too little.

One never learns anything while talking; but always is in the way of learning by listening when others talk.

53. The habit of accepting favours from others without reciprocating.
54. Intentional dishonesty in business and professional relations.

Note: Check yourself carefully by these fifty-four causes of failure; if you can check yourself O.K. after each of these causes of failure you need not worry about any fear and you have everything under control.

After you have made your own rating, however, it may be both interesting and useful if you will have some other person (your friend or a relative) rate you on each of these causes of failure—someone who knows you quite well and has courage to let you look at yourself through his or her eyes.

24

Academic Terms

Academic Board/Committee

The higher academic body within the University system responsible for all academic standards and related policies and practices.

Academic Misconduct

Conduct in relation to any academic work that is dishonest or unfair, and includes all forms of cheating, plagiarism, and lying.

Academic Record

The complete academic history of a student at the University. It includes personal details, courses completed, assessment results (marks and grades), and any awards or prizes won by the student.

Academic Staff

Faculty/staff who has been appointed to undertake 'teaching and research' activity. Assistant Professors/Associate Professors/ Professors and Deans are considered as Academic Staff/Faculty.

Academic Year

The time during the year when teaching normally takes place, generally from June/July to May (next year). The academic year consists of two semesters; sometimes courses are offered during Summer Term also towards the end of the Academic Year.

Administration

The implementation of authorised policies and procedures, and the application of systems developed within the University to achieve the agreed results and performance.

Assessment

The process by which a student's academic progress or achievement in a course/programme is evaluated. This includes assignments, quiz, submission of term-end papers, class tests, project work, seminars, tutorial participation, and final examinations. Assessments in the performance and creative arts can include studio practice, modelling, creative productions, paintings, production of video and films, and artistic performances and media work. Concerned faculty is responsible for evaluating the work done by student and assigning a grade of assessment to each enrolled student at the completion of the course.

Assessment needs to be aligned with the learning outcomes of the course/programme. It should be valid, reliable, and consistent with the development and demonstration of skills and knowledge acquired in the professional context.

Assessment may be formative, providing learners with a feedback on their progress during the semester, or summative, providing an overall measure of achievement of a learner's performance in relation to the learning outcomes of the course as a whole at the end of the semester/year.

Assessment Grades

Examination's result in a course is normally linked with a % range of marks. Student's achievement in each course is assigned a grade of assessment. The following letter grades are normally used.

Grade	Description	Marks (%)
D	Distinction	80 to 100%
A+	Very Good	70 to 79%

Grade	Description	Marks (%)
A	Good	60 to 69%
B	Pass	50 to 59%
F	Fail	Less than 50%

Assignment

An academic work submitted usually by an individual student based on the learning that occurs in class and the determination of how well the students have learnt. Although, most assignments are individual, some are group assignments, submitted by a group of students on a particular topic.

Audit Agency (Quality Audit Agency)

An independent, national/international agency that conducts quality audits of universities, colleges, and other higher education institutions of State and Central Universities and provides relevant Certification valid for a definite period of time (years). Such auditing and certification can be for a particular programme also in a University set-up.

Authentication

An IT (Information Technology) term related to checking of permission granted with a user ID and password when an authorised user logs on to a computer system.

Award (of Degree)

A recognised Certification of achievement or competency. Degree certificate is issued to a student after satisfactory completion of all the requirements of a programme/degree. The term Award, sometimes, is also used to describe a scholarship/ stipend given to support the study in a University course.

Bachelor's Degree

A University Degree usually requiring 3 or 4 years of fulltime or equivalent part-time study. A Bachelor's Degree qualifies a person for entry into service, occupation, or a profession.

Bench Marking

A process of systematically measuring and comparing the procedures/processes, products, services of an organisation/ industry with those of recognised leading organisations/industry with the purpose of improving its own performance.

Board of Examiners

The Committee of Faculty members that examines and approves the individual student's grades and marks in the courses and determines the academic status of each student, and also determines the students who are to be given, Awards, Prizes, and other Honours. Usually, a Dean chairs the Board of Examiners.

Campus

The physical location of the University/College from where a programme/course is offered. In addition to the main campus there may be campuses in other locations/countries/cities where students may be enrolled.

For online or other distance education courses, the main campus is the location where the course material is developed, maintained, and delivered.

Chancellor

The Chief Executive Officer/Head of the University. The Chancellor chairs the meetings of the University Council and presides over graduation ceremonies. In some Universities, the term Provost, or President is used analogus to Chancellor.

Cheating

Obtaining or attempting to obtain, or helping in obtaining, credit for academic work through any dishonest, deceptive, or fraudulent means. Cheating includes copying or using unauthorised information/sources, materials, or data during a test or an examination, and the fabrication or falsification of such data.

Contact Time

The duration in a course where students are required to attend classes/periods taken by a faculty during lectures, tutorials, seminars, laboratories, and workshops. Expressed in hours of instruction in a semester.

Council (Academic Council)

The governing body that controls and manages the administration, operations, and academic affairs of the University. University Academic Council consists of members of the academic staffs drawn from different Departments/Schools. Meeting of the Academic Council is chaired by the Chancellor/ President, or by Vice-Chancellor.

Course

A programme of study necessary to qualify for an award at the University. Courses are usually identified by a course code (alpha-numeric). Some of the courses may be foundation courses, prerequisite courses, and some may be core courses/ elective courses and advanced courses.

Course Code

A sequence of usually three characters and three numbers (alpha-numeric code) used to identify each course of the University, to identify whether it is a UG or PG course, the programme, the year of study, and whether foundation, core/ elective or advanced course.

Course Completion

The satisfactory completion of all the academic requirements of a course including any mandatory minimum attendance, assignments, quizes, examinations, assessments, dissertations, practicals, and work experience/project work in industry. This enables the student to become qualified for the conferring of the Award of a Degree/Diploma at graduation ceremony.

Course Coordinator

A senior faculty responsible for the conduct and coordination of the course. The Course Coordinator is responsible to the Head of School/Department for ensuring that each course meets its learning outcomes as approved by the Academic Board, the needs of the learners, conduct of examinations, and the expectations of the professional community. Also called as Programme Coordinator.

Course Learning Outcomes

The learning outcomes of the course stated clearly and in terms that enable the students to understand what skills and knowledge they are expected to achieve, and what values and attitudes will be developed by satisfactorily completing the course. The learning outcomes of the unit are usually expressed as: "On completion of this course students should be able to:" followed by the list of specific learning outcomes. The course learning outcomes are included in the curriculum outline/ document.

Course Outline

A statement of the details of the course in terms of different units and how it is taught. The course outline usually contains the course title, prerequisites, credit points, learning outcomes, content, teaching method, exercises to be completed, assessment, texts, and references. The course outline is a document that is approved by the School/University, providing a clear statement about the course to both the students and faculty.

Course Plan

A plan of how the course is being offered and run in a given teaching period. A course plan is a supplement to the course outline available in the curriculum document. It indicates how the course is being offered and run in the given teaching period (e.g., semester). It contains the schedule of topics/lectures, dates for quiz, submission of assignments, and end-semester examinations.

Course Structure

The organisation of the programme in terms of courses/ subjects to be undertaken by the students in each Semester/ Year, the study and evaluation scheme, the learning outcomes, and other mandatory requirements for the Award of the Degree.

Dean

The academic head with responsibility for guiding/assisting the faculty, strategic and policy issues, and procedures and practices related to teaching and learning, research and other academic work.

Discipline (or Branch)

A field of study (or knowledge) consisting of a related group of courses/subjects. Such fields of study/knowledge may be unitary, inter-disciplinary, cross-disciplinary, multi-disciplinary, and/or trans-disciplinary.

Distance Mode

An off-campus form of course delivery. Distance mode of delivery of a course usually refers to printed materials (study guides, work sheets, work books, assignments) being sent to individual student by mail/e-mail, and students completing the exercises and sending them back to the Faculty/University, for evaluation and grading. Sometimes, the study materials may also be in the form of MMLP's (Multi-Media Learning Packages).

Doctoral Degree (Doctorate)

Highest University award (Ph.D., D. Sc., D. Ed., or Dr. Eng.) given by the University. This Degree is awarded for a substantial and original research and scholarly work which has been carried out under the guidance and supervision of a Professor/a Guide for a period of usually 3 years full-time or equivalent part-time study.

Due Date

The date (and possibly the time of the day) when an assignment is due to be submitted. The due date for each assignment is listed on the course plan/assignment sheet and date of submission of assignment is written on the front page of the assignment sheet submitted by the student. Submission of an assignment after the due date may result in a lesser grade/mark.

Electives

Approved courses selected from the range of courses offered within the University Programme/Curriculum to make up the number of credits necessary to complete a Degree. A programme may make provision for elective courses to enable students to broaden their horizons of knowledge to suit their personal interest areas/specializations.

Enabling Course/Bridge Course

A programme of study which provides bridging or supplementary education for the purpose of enabling a person to undertake programme at a university. Any deficiency in the requirements for entry can be made up by such preparation courses.

Examination

The formal end-of-semester assessment in a course conducted during the scheduled two-weeks examination period. Although most of the examinations are written, but some are practical examinations. Students provide written answers to questions asked in the examination paper. These examinations are usually of two or three hours' duration and usually of closed books, sometimes may be open book.

In an University set-up, Controller of Examinations plans and conducts all the examinations centrally, including their supervision and evaluation of answer scripts.

External Candidate

Usually refers to a Doctoral Scholar who is working full-time in an organization and has registered for a Ph.D. Programme in the University on a part-time basis.

Faculty

Generally refers to a teacher/academic staff working in the College/University. Sometimes the term may refer to an academic unit in the University, e.g., Faculty of Arts, Faculty of Engineering, Faculty of Sciences, Faculty of Management, etc.

Faculty Adviser

A faculty member, working in a School/Department, assigned the responsibility of advising and counselling students. He meets students in one-to-one or a group of students may contact the Advisor either personally or by e-mail/phone or by fax.

Foundation Courses

A set of courses that are common to a range of programmes, e.g., first year courses for which almost every student is required to register and pass.

General/Administrative Staff

University staff engaged in administrative and support services, and who are not engaged in the academic tasks of teaching and/or research.

Graduand

A student who has completed all the requirements for the award of a Degree and is waiting to get the Degree Certificate in a Graduation Ceremony/Convocation.

Graduate

A person who has received a Degree after satisfactorily completing all the requirements of the programme. The term "Graduate" is also used to describe a programme such as a graduate programme in Mechanical Engineering.

Graduate Attributes

Generic personal attributes, skills, capacities and capabilities, and values relevant for employment, that help make the graduate a valuable and worthy employee. Department/School develops graduate attributes and learning outcomes/goals specific to their disciplines/programmes/studies.

Head of the Department (HOD)

The senior academic with the responsibility entrusted for all the activities in the Department/School. Each School/ Department has a head for managing all the activities related to teaching and learning within the School/Department and for carrying out R&D.

Handbook (University Handbook/Calendar)

A University publication containing essential information for all students, faculty, and staff on the University facilities, procedures, rules and regulations, courses, and schedule of important dates as per the Academic Calendar drawn in advance. Usually published in two Volumes, an Undergraduate Handbook and a Post-Graduate Handbook.

Internationalisation

The process of collaborating with international institutions through an MoU (Memorandum of Understanding) for teaching, research, and for carrying out joint projects for mutual benefits.

Internet

The computer communication system created for the interconnecting networks of computers around the world for exchange of informations and for sending/receiving e-mails.

Key Performance Indicators (KPI)

A measure of key performance for an organisation or a functional unit within the organization. It is an indicator of a factor critical to the success of the organization.

In respect of a faculty, KPI are: extent of teaching excellence, research accomplishments/publications, projects, co-curricular, and extra-curricular activities carried out.

Learning

Improving and increasing one's knowledge of and skills in a discipline by study through instruction or by experience. Learning occurs at individual level.

Learning Approach

How and why the student learns? A learning approach has two components—a motive (why the student wants to learn) and a learning style (how the student learns). Two commonly recognised approaches to learning are the surface approach, i.e. reading by glancing and the deep approach (in-depth study).

Lecture

A formal presentation by a faculty member, usually to a large number of students in a lecture class. Lectures may include the use of course material, handouts, PPT/audio-visual presentations/aids. There are usually few opportunities for faculty-student interactions, and questioning during a lecture. However, trained faculty members encourage their students to discuss and ask questions and clarify their doubts.

Management for Performance

Performances of faculties and review of their academic works are carried out annually. The system provides an opportunity for faculties to discuss their performances and future goals with their supervisor/Head of School or Department to align individual performance with organizational objectives. For new faculty, the system provides for special faculty development programmes (FDP's) to clarify his/her work expectations.

% of Marks

Performance of students is indicated as a % marks obtained in a course. A range of % marks is linked to a Letter Grade.

Marking Scheme

A guideline outlining how the marks for an assessment or examination answer scripts are allocated enabling different examiners (faculty) to assess students' works fairly and consistently. The marking scheme for assessment in a subject is usually prepared by a faculty coordinator. In respect of end-semester examinations, the examiner provides a marking scheme.

Masters Degree

A University degree designed to provide advanced theoretical and applied studies in a selected discipline. A Masters Degree usually requires an advanced study of upto two years full-time following completion of an U.G. Degree/Bachelors Degree. A Masters Degree is usually classified as a masters Degree by Coursework and a Thesis or Masters Degree entirely by Research.

Mode of Study

A classification of the manner in which a student is undertaking a degree programme/course. Modes of study in respect of programmes are full-time, part-time (internal), or external. Sometimes the courses may be by self-study or guided study.

Modularisation

A flexible course structure in which the entire course and its units are structured around a number of independent or sequenced modules. Each individual module has a specified course content and assignments to be submitted for evaluation and grading. Some flexibility is offered in completing the modules required depending on the pace of the learner, his/her prior knowledge, and interest/individual needs.

Orientation Programme

A range of programmes and activities designed to assist first year students for smooth entry into University system. The University schedules orientation in the week immediately prior to the beginning of the First Semester Classes.

Plagiarism

The intentional or unintentional presentation as one's own work, the work of another student/faculty/author without appropriate acknowledgement or references. The term 'work' includes all or part of any form of work including notes and papers, assignments, seminar and conference papers, reports, theses, articles, books, lecture notes, and electronic submissions. Plagiarism includes:

- copying without acknowledgement of the source or giving references,
- copying the text of another student's assignment,

- copying of computer programmes, and
- copying of visual representations (line drawings, diagrams, photos).

Policies

The rules which govern student's enrollment and their academic progress in the University. Policies provide a clear statement of the University's regulations on academic and student issues, and the procedures to be followed.

Postgraduate

Refers to a programme or a student; a postgraduate course is either a doctoral/master or a postgraduate diploma course. In U.S. Universities, postgraduate programmes are usually referred to as Graduate programmes. A Postgraduate student is one who is enrolled in a postgraduate course after his undergraduate study.

Practicum

A part of a course providing learning experiences outside the University in a relevant professional setting such as an industry/ organization or Research Laboratory. During the professional placement, the student may either be an active participant or merely an observer. A practicum may cover ether part of a course or many courses within the curriculum.

Prerequisite Course

A requirement that must be satisfied prior to enrollment in a specific course. A prerequisite course is usually a pass in another course (or an exemption gained from a pass in an equivalent course). Not all courses have prerequisites. All matters related to prerequisite are the responsibility of the Course/Programme Coordinator.

Programme/Degree

A field of knowledge/discipline or study comprising a related group of courses, areas of research, or professional practices.

Pro Vice-Chancellor

A senior member of the faculty or a Dean with a specific responsibility assigned to help the Vice-Chancellor in academic and related issues. Pro Vice-Chancellor may also hold position as a Dean.

Quality

Fitness for purpose, satisfying the requirements of objectives and specifications. Quality is related to standards meeting the specifications and the levels of attainment required.

School/Department

An academic unit comprising of Faculty, academic facilities, and laboratories pertaining to a discipline. The School in headed by a Dean/HoD.

Semester

An academic teaching period. Normally each semester consists of 16 weeks of instruction and a final assessment/ examination period (usually of two weeks). In each academic year, there are two semesters—odd semester (usually between June/July and December) and even semester (between January and May/June).

Seminar

A seminar incorporates a lecture format combined with a discussion to follow where the presenter can answer questions or further explain the material presented. A seminar usually involves a “small group” rather than a “large lecture group”. A seminar is well suited to an audience where experts are also in the audience and they can then complement the presentation with their own experience. A seminar is usually presented in a class/Lecture hall where the audience is in close proximity to the presenter.

Student

A learner for whom there is an enrollment in a University course at the beginning of the Semester. A student can pursue his studies on a part-time basis or on a full-time basis.

Student ID

The identification card or badge issued by the University to all students on enrollment. Each student ID may contain a code of letters and numbers allocated sequentially.

Student ID Card

The identification card issued for all students containing the student's name, his photograph, roll number, name of the programme/course, and a library bar code.

Student/Staff Ratio

The ratio between the number of full-time students taught by the Department and the number of full-time faculties in the Department.

Summer Session

A special teaching/instructional period during the summer vacation and is of a short duration, usually 6 weeks; Summer session provides an opportunity to accelerate the course work, upgrade professional knowledge and/or to extend the workplace skills and employment prospects.

Teacher

An academic staff member, faculty, lecturer, or professor involved in the teaching of a course.

Teaching and Learning Portfolio

A written, reflective document, summarising teacher's approach to teaching and learning, in respect of a particular course or subject and providing evidence of methods used, achievements in teaching, and teaching/academic awards, and accomplishments.

Timetable

A schedule of all courses offered in a given semester showing the contact times (from .... to ....) and room locations for all classes (lectures, tutorials, seminars, laboratories, and workshops).

Tutorial

An instructional activity carried out by a faculty member, with a small number of students, structured to enable students to develop specific skills and knowledge under faculty supervision. The tutorial provides opportunities for effective staff-student interaction, questioning and clarifying the difficult concepts taught in the lecture classes. Individual guidance is given to the students in solving difficult numerical problems.

Undergraduate

The term refers to a course/programme or student category. A course leading to an award of a Bachelor's Degree. An undergraduate student is one who is enrolled in an undergraduate course/programme.

Vice-Chancellor

The chief academic head of the University, responsible for various academic functions like teaching, R&D, consultancy and extension services, and community development at the university level.

World Wide Web

The body of software, protocols, and conventions that make the internet easy for anyone to browse and use.

Appendices

APPENDIX I

Industrial Training Programme

General Instructions for Students

- 'Learning New Things' and 'Gaining Practical Experience' are the key objectives of industrial training programme.
- Always have a Diary with you during the training period.
- Follow the Training Schedule given by the industry meticulously.
- 'Be on time' everyday for training.
- Carry out the assignments given by the industry daily, with all sincerity.
- Be keen to observe the new features followed in Design, Material, Production, Administration, etc., in the industry.
- Make a note of such findings in the Diary, on the same day. Don't postpone.
- In case of any doubt arising in the industrial processes, do not hesitate to get it clarified from the engineer/ supervisor in the industry.
- Do not mind soiling your hands and dress. Hands on experience will give greater confidence in you.

- Meet as many engineers as possible and engage in technical discussions with them. This will make you a confident professional.
- Make friends, continue the friendship for mutual help at a future date.
- Look for opportunities for taking initiatives, shouldering responsibilities, leading a team of employees on an assignment, projecting your innovative ideas for improvement, etc.
- Identify the unique features of the industry in various aspects and note down in the Diary.
- 'Managing Men' is key issue for becoming a successful engineer/manager. Learn the best practices followed in the industry and make a note of them in the Industrial Training Diary.
- Follow the rules and regulations of the industry, where you are undergoing industrial training.

Industrial Training Diary

Day/Week: 1/1 Date: ....

Company Name: ....

Division/Section: ....

Notes: Write here what you have observed in the industry (or) what new knowledge or information you have obtained.

....

....

....

....

....

....

....

....

....

Signature:
Date:

CERTIFICATE

(Issued by the Company/Industry)

Certified that the candidate Mr./Ms. ____________________ studying in ____________________________________ Branch at ____________________________ Engineering College has undergone the prescribed industrial training in our company during _________________ to _________________. He/She has completed the industrial training satisfactorily as per the Training Schedule.

Signature with date
(Officer Incharge/Industry Guide)

Company Seal

Countersigned by Industry Training Coordinator of the Department in the Engineering College.

Date:

APPENDIX II

Project Work Diary

Name of Engineering College: ....
Address: ....
Name of Student: Mr./Ms. ....
Regn. No./Roll No.: ....
Branch: ....
Project Title: ....
Name of the Guide: ....
Organization/Industry where Project Work is being carried out (with address): ....

....

....

....

....

Details to be filled in only when project work is carried outside the College

1. Name of the Organization/Industry: ....
2. Address of the Organization: ....

....

....PIN..........................

3. Telephone Number with STD Code: ....
4. Fax: ....
5. E-Mail: ....
6. Main Activity of the Organization/Industry: ....

   Products Manufactured: ....
7. Name of the Industry Guide: ....
8. Name of Faculty Guide: ....
9. Designation: ....

10. Office Phone No.: ....

11. Intended Period of Study: ....

12. Facilities Available for Students: ....

....

13. Timings to be Followed During Project Work:

....

14. How to reach the Organization?

....

15. Brief Description of the Intended Project:

....

....

....

....

Date: Signature of the Student

Signature of the External Guide (If the Project Work is carried out in an Industry) Signature of the Faculty Guide

---

1. Introduction

Project Work is an important component of B.E./B.Tech Degree programme. In undergraduate level, in the final semester, all students are required to carry out a Project Work either individually or in groups of two to five students. Projects are carried out by the students under the guidance of a teaching/ industry faculty. The projects could be carried out within the campus or outside the campus according to the nature of the project. If carried out outside the campus, an external/industry guide is also needed to supervise and monitor the progress of the project work.

The primary objective of a project work is to give an exposure to the students on the application of concepts and theories that they have learnt in the course. The projects can be an analysis of existing work practices or design and development of new processes, procedures, and equipment or a working model or a system to solve a real life problem. The selection of project area depends on many factors including facilities available, requirements of the industry, time and interest of the students. An excellent project work can get not only good grades but also better job offers from industries.

2. Objectives

Some specific objectives of final year students' projects are:

- To expose the students to industrial environment and work practices where manufacturing activities are carried out so that they are mentally and physically prepared to serve in an organization after their graduation.
- To enable the students to observe and learn from the industry experts or practicing engineers and professionals.
- To enhance the communication skills of the students by providing opportunities to discuss in groups and to present their observations, findings, and report in formal reviews both in oral and written format.
- To identify the relevant industry practices when compared to what are being taught in the institutional/laboratory set-up.
- To know the latest industrial trends and improve the technical expertise of the student.

Some specific problem oriented projects can also give opportunities to the students to offer new and innovative solutions in the form of suggestions or alternatives for consideration by industry/organization for implementation.

3. Project Selection

In the Engineering College, an introduction to project work covering need, importance, topics/areas available for study, prospective places/industry for doing Project Work, and methodology are provided during the sixth semester. Sixth semester students attend the presentations made by eighth semester students on their project work. In the first week of eighth semester the students are grouped by the Department/Class Coordinator or on their own and then the student groups identify their study area or the project that they would like to take.

Faculty guides are assigned based on their specialization and interest. Final year Project Work Coordinator in consultation with students finalises the student groups and guides, and the same is communicated to them. (List of students in the group, name of the guide, topics of project work [tentative], and name of the company/industry in which the project work will be carried out [if, any], will be displayed on the notice board.)

Once this process is over, the faculty guide and the team will discuss and explore the places of project work and actual title of the project work. They are expected to furnish a Preliminary Report concerning Project Work (Proforma 8) to the Project Work Coordinator within two weeks from the commencement of eighth semester.

4. Project Schedule

Each student/team is expected to prepare a Project Schedule as per the format given in Proforma 9. (Project Schedule: Tentative Plan). It will be updated as and when each activity is carried out. Effort should be made to adhere to the planned schedule. If any major deviation is noticed, then a suitable reschedule may be worked out in consultation with the Guide.

5. Project Supervision

1. The Project Schedule prepared by the students' shall serve as a base for guidance and supervision of the project and its progress.
2. As per the Schedule project review meetings will be conducted periodically and recorded in the Project Work Diary.
3. If a project work carried out in the college requires visits to industry/organizations by students it should be recommended by the Guide. After visiting the

industry the students should apprise the Guide about the discussion and progress made during the visit.

4. The student is expected to submit three Project Review Reports (Proforma 11: Project Review Report) to the Project Coordinator as per the Schedule. It should be signed by the student.
5. A draft version of the Report will be prepared and presented to the Guide. The observations and suggestions made shall be considered and carried out, and final version of the report is submitted to the Department.
6. The Guide can explore the possibility of getting assistance to support the project from various State and National funding agencies.
7. The Guide will plan and get necessary permission and approval from the Dean/HoD for any special device required and ensure that facilities needed for the completion of the project is available.
8. It will be the responsibility of the Guide to ensure that the project is useful to the student and the industry. Students can be encouraged to develop a new system, new model, carry out a small scale research and publication of a paper. Faculty Guide should ensure that the project work could be completed well within the time frame available.
9. If a project work is carried out by a group of students, the tasks assigned and carried out by individual members towards the progress of the project will also be recorded in the Proforma 18.
10. At the end of the project, student is expected to give individual feedback about the project work and the benefits derived (Proforma 19).

6. Outcome of the Project

One or more of the following are generally expected as outcomes of the Project Work:

- Innovative solution to the selected problem.

- Development of new design/circuit/product/process/layout/procedure/model.
- Critical analysis of existing practices and offering suggestions for improvements.
- New software development/development of a new algorithm.
- Publication of a paper in a technical journal/presentation of a paper in a conference.
- Application and registration of a patent.

7. Project Evaluation and Scheme of Assessment

Suggested criteria for assessment of the project work are:

Continuous/Semester Assessment			Final Examination	
S. No.	Criteria	Maximum Marks	Criteria	Maximum Marks
1.	Literature Survey and Problem Definition	20	Presentation and Viva-voce	50
2.	Methodology and Plan of work (I Review)	20	Project Report	50
3.	Word carried out (II Review)	20		
4.	Results and Analysis (III Review)	20		
5.	Final Report	20		
Total		100	Total	100

8. Preliminary Report Concerning the Project Work

Name of the Student: Mr./Ms. ....

Registration No./Roll No.: ....

Branch: ....

Faculty Guide's Name: ....

or/and Industry Guide's Name: ....

Broad Area of Study: ....

....

Place where Project Work will be carried out: ....

....

Specific Problem Identified/Proposed Project Title: ....

....

....

Date: Signature of the Student

Signature of the Industry Guide (If the Project Work is carried out in an Industry)

Signature of the Faculty Guide

9. Project Schedule [Tentative Plan]

Name of the Student: Mr./Ms. ....

Branch of Study: ....

Faculty Guide's Name: ....

Project Title: ....

[Identify main activities in your project work and write down here in a chronological order with its target date.]

Major Activities	Target Date
Preliminary Report concerning the Project Work	
Finalisation of Objectives and Methodology	
Review-I	

Major Activities	Target Date
Review-II	
Review-III	
Submission of draft version of Final Report	
Presentation and Viva-voce	

Date: Signature of the Student

Signature of the Guide

10. Record of Activities Carried out by the Student in Month-1

Month: .... Year: ....

[You may also record here the points discussed.]

Week	Activities carried out during the week
1st Week	
2nd Week	
3rd Week	
4th Week	

Signature of Faculty/Industry Guide:

Name of the Guide:

Date:

11. Project Review Report-1

Name of the Student: Mr./Ms. ....

Registration No./Roll No.: ....

Branch of Study: ....

Faculty Guide's Name/Industry Guide's Name: ....

....

Project Title: ....

Status of the Project: ....

....

Date: Signature of the Student

Remarks by Guide: ....

Signature of the Guide with date

Project Review Report should be submitted during reviews. This report shall give the details of work carried out till now and progress of the project work.

---

12. Record of Activities Carried out by the Student in Month-2

Month: .... Year: ....

[You may also record here the points discussed.]

Week	Activities carried out during the week
1st Week	
2nd Week	
3rd Week	
4th Week	

Signature of Faculty/Industry Guide:

Name of the Guide:

Date:

13. Project Review Report-2

Name of the Student: Mr./Ms. ....

Registration No./Roll No.: ....

Branch of Study: ....

Faculty Guide's Name/Industry Guide's Name: ....

....

Project Title: ....

Status of the Project: ....

....

....

Date: Signature of the Student

Remarks by Guide: ....

Signature of the Guide with date

Project Review Report should be submitted during reviews. This report shall give the details of work carried out till now and progress of the project work.

---

14. Record of Activities Carried out by the Student in Month-3

Month: .... Year: ....

[You may also record here the points discussed.]

Week	Activities carried out during the week
1st Week	
2nd Week	
3rd Week	
4th Week	

Signature of Faculty/Industry Guide:
Name of the Guide:
Date:

15. Project Review Report-3

Name of the Student: Mr./Ms. ....

Registration No./Roll No.: ....

Branch of Study: ....

Faculty Guide's Name/Industry Guide's Name: ....

....

Project Title: ....

Status of the Project: ....

....

....

Date: Signature of the Student

Remarks by Guide: ....

Signature of the Guide with date

Project Review Report should be submitted during reviews. This report shall give the details of work carried out till now and progress of the project work.

---

16. A record of Interaction/Discussions between Student and Faculty/Industry Guide during the Project Work

Date	Points Discussed/Action Proposed

Signature of Faculty/Industry Guide:
Name of the Guide:
Date:

17. Record of Challenges Faced by Student(s) during the Project Work and How was it Solved

Project Title: ....

....

[Record here major hurdles/problems faced by you/your team in completing the project work. Briefly discuss what steps have been taken to overcome or to solve the problem, if any.]

S. No.	Major Problems/Challenges	How was it solved?

18. Project Work Undertaken by a Group of Students

If a project work is carried out by a group of students, list out the tasks assigned and carried out by each member. This information is required to know the specific contribution made by each member of the group in carrying out the project work.

Name of the Members in the Group	Tasks Assigned/Major Work carried out
Your name	
Names of the other members in the group	
Name (1)	
Name (2)	
Name (3)	

Guide's Remarks:

Signature of Guide:
Date:

19. Student's Feedback

Name of the Student: Mr./Ms. ....

Registration No./Roll No.: ....

Branch of Study: ....

Faculty/Industry Guide's Name: ....

....

Project Title: ....

....

....

Name of the Organization: ....

Write your feedback in respect of Project Work you have carried out/benefits you have derived.

....

....

....

....

APPENDIX III

Internet Information Sources

Today internet and its related technologies like www. have resulted in the rapid proliferation of electronic information. Information resources available on internet are:

1. **google.com** Google enables users to search the information, including web pages, images, and videos. It offers unique features and search technology.
2. **yahoo.com** is a major internet portal and service provider offering search results, customised content, chat-rooms, free e-mail, clubs, and pager.

APPENDIX IV

Dimensions

Dimensions are used to describe physical properties of objects and materials used in engineering.

There are Fundamental Dimensions and combinations of such dimensions which are called Derived Dimensions.

Examples of Fundamental and Derived Dimensions

Fundamental Dimensions:

- Length, L
- Time, T
- Mass, M
- Electric current, I
- Temperature
- Amount of substance (mole)
- Luminance intensity

Derived Dimensions:

- Area: L^2
- Volume: L^3
- Velocity: $\frac{L}{T}$
- Acceleration: $\frac{L}{T^2}$
- Mass density: $\frac{M}{L^3}$
- Force: $\frac{M.L}{T^2}$
- Energy: $\frac{ML^2}{T^2}$

Some quantities used in engineering calculations have no dimensions. These are called dimensionless quantities. Dimensionless quantities are ratios of quantities having the same dimensions; e.g., Π (ratio of circumference of a circle to its diameter).

Units; Système Internationale d'unitès (SI) (Since 1960)

Table 1: Seven Base Units of SI

S. No.	Quantity	Unit	Symbol
1.	Length	Metre	m
2.	Mass	Kilogram	kg

S. No.	Quantity	Unit	Symbol
3.	Time	Second	s
4.	Electric current	Ampere	A
5.	Thermodynamic temperature	Kelvin	K
6.	Amount of substance	Mole	mol
7.	Luminous intensity	Candela	cd

Table 2: Supplementary Units (neither "Base" nor "Derived")

S. No.	Quantity	Unit	Symbol
1.	Plane angle	Radian	rad
2.	Solid angle	Steradian	sr

Table 3: Derived Units

Quantity	Unit	Symbol	Formula
Frequency	Hertz	Hz	s^{-1}
Force	Newton	N	$kg.m/s^2$
Pressure, stress	Pascal	Pa	N/m^2
Energy, work	Joule	J	N.m
Power	Watt	W	J/s
Quantity of electricity	Coulomb	C	A.s
Electric potential	Volt	V	W/A
Capacitance	Farad	F	C/V
Electric resistance	Ohm	Ω	V/A
Conductance	Siemens	S	A/V
Magnetic flux	Weber	Wb	V.s
Magnetic flux density	Tesla	T	Wb/m^2
Inductance	Henry	H	Wb/A
Luminous flux	Lumen	lm	cd.sr
Illuminance	Lux	lx	lm/m^2

Table 4: Other Derived Units

Quantity	Unit	Symbol
Area	Square metre	m^2
Volume	Cubic metre	m^3
Velocity	Metre per second	m/s
Acceleration	Metre per second squared	m/s^2

Quantity	Unit	Symbol
Density	Kilogram per cubic metre	kg/m^3
Specific volume	Cubic metre per kilogram	m^3/kg
Entropy	Joule per kelvin	J/K
Radiant intensity	Watt per steradian	W/sr
Bending moment, torque	Newton-metre	N.m
Heat capacity	Joule per kilogram-kelvin	J/kg.K

Table 5: List of Non-SI Units

Quantity	Unit	Symbol	Meaning
Time	Hour	h	3.6 ks
	Minute	min	60 s
	Day	d	calendar
	Week	. . .	calendar
	Year	. . .	calendar
Angle	Degree	°	$(\pi/180)$ rad
Volume	Litre	l	$1\ dm^3$
Temperature	Degree Celsius	°C	$t_c = t_k - 273.15$

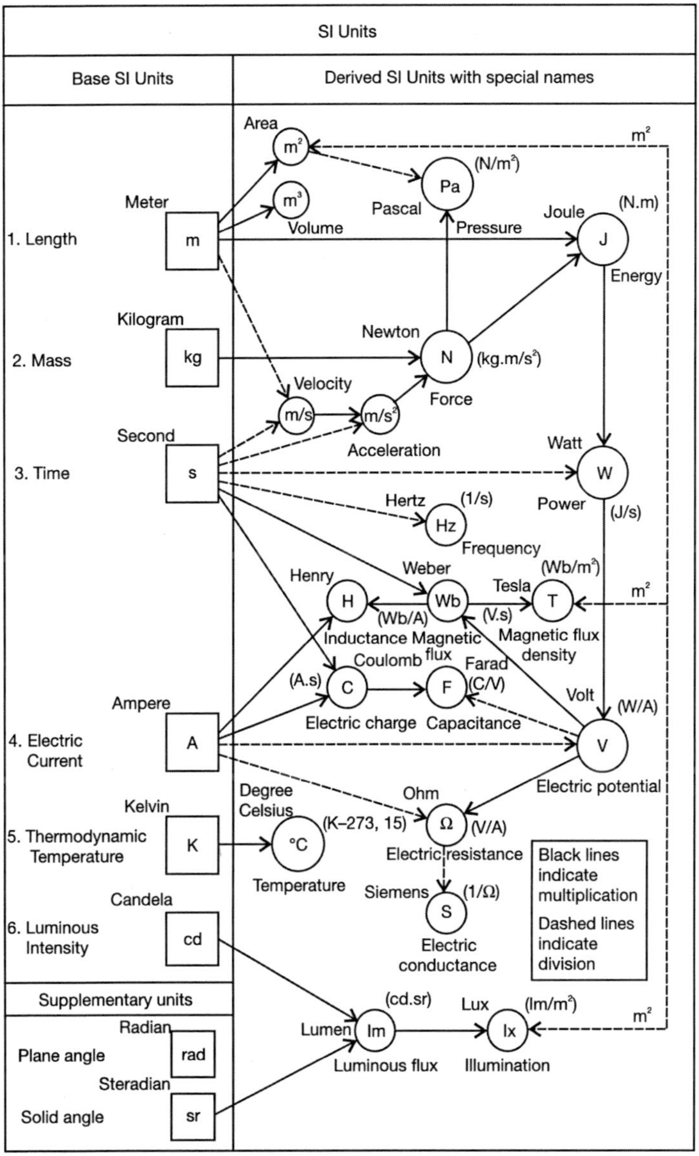

Fig. 1: Relationship between Physical quantities and SI units

Table 6: SI Prefixes

Multiplication Factor	Prefix	Symbol
10^{12}	tera	T
10^{9}	giga	G
10^{6}	mega	M
10^{3}	kilo	k
10^{2}	hecto	h
10^{1}	deca	da
10^{-1}	deci	d
10^{-2}	centi	c
10^{-3}	milli	m
10^{-6}	micro	μ
10^{-9}	nano	n
10^{-12}	pico	p
10^{-15}	femto	f
10^{-18}	atto	a

Table 7: The Greek Alphabet

Α α	Alpha	Ν ν	Nu
Β β	Beta	Ξ ξ	Xi
Γ γ	Gamma	Ο ο	Omicron
Δ δ	Delta	Π π	Pi
Ε ε	Epsilon	Ρ ρ	Rho
Ζ ζ	Zeta	Σ σ	Sigma
Ε η	Eta	Τ τ	Tau
Θ θ ϑ	Theta	ϒ υ	Upsilon
Ι ι	Iota	Φ φ	Phi
Κ κ	Kappa	Χ χ	Chi
Λ λ	Lambda	Ψ ψ	Psi
Μ μ	Mu	Ω ω	Omega

Gravitational System and Absolute System

In a Gravitational System, the units of mechanics are all derived from three fundamentals, viz., length, force, and time; and the unit of mass is a derived unit.

In an Absolute System, the units of mechanics are all derived from three fundamentals, viz. length, mass, and time; and the unit of force is a derived unit.

In either of the two systems, the units of mass and force are always related by *Newton's Law; F = ma.*

Table 8: World's Measurement Systems

	English System	Metric System
Gravitational System		
mass	Slug (lbf.s^2/ft)	kg (kgf.s^2/m)
force	pound-force	kilogram-force
Absolute System		
mass	pound	kilogram
force	poundal (lb.ft/s^2)	newton (kg.m/s^2)

APPENDIX-V

Important Conversion Factors

1 inch = 2.54 cm = 25.4 mm

1 yard = 3 ft. (1 ft = 12 inches) = 0.914 m

1 meter = 39.37 inches = 3.281 feet

1 sq. inch = 6.45 Sq. cm

1 sq. meter = 10.76 Sq. ft

1 gallon = 4.54 litres

1 U.S. gallon = 3.785 litres

1 litre = 1000 cm^3 = 1000 cc

1 pound (lb) = 0.454 kg

1 kg = 2.205 lb. (pounds)

1 mile = 1.6 km

1 acre = 4050 sq. metres = 0.405 hectare = 4840 sq. yard

1 hectare = 10,000 sq. metres = 2.5 acres

1 ounce = 28.35 gms

1 ton = 1000 kg

1 HP = 0.746 kW

1 stone = 14 pounds

$$\text{Temp.}^\circ\text{C} = \frac{(\text{Temp.}^\circ\text{F} - 32)}{1.8} \quad (^\circ\text{C} = \text{Degrees Celsius})$$

$$\text{Temp.}^\circ\text{F} = (\text{Temp.}^\circ\text{C} \times 1.8) + 32 \quad (^\circ\text{F} = \text{Degrees Fahrenheit})$$

1 μm = 1 micron = $\frac{1}{1000}$ mm = 10^{-6}m

1 μm = 1 micron = 40 μ inch (micro inches)

1 unit of energy = 1 kW hr. = 3.6 MJ.

1 Watt = 1 Joule/sec. = 1 Nm/sec

1 Joule = 1 Nm (Newton-metre) = 1 Watt-second

1 Newton = 1 kg m/sec^2

1 kg weight (kgf) = 9.81 Newtons

1 atmosphere = 1 kg/cm^2

1 kg/cm^2 = 15 psi (pounds per sq. inch).

1 bar (unit of pressure) = 10 N/cm^2 = 0.1 MPa

= 15 psi (approx. 1 atmosphere pressure)

1 radian = 57.27 degrees

1 Pascal (Pa) (unit of pressure) = $\frac{N}{m^2}$

1 nautical mile = 1.8 km

1 geographical mile = 1.6 km

1 carat (diamond) = 0.2 gram

1 sovereign (gold) = 8 grams

1 hundred weight = 112 pounds

1 m/sec = 3.6 km/hr

1 r.p.m = 0.105 radian/sec

1 mach = 350 m/sec

1 barrel = 120 to 160 litres (oil industry)

1 cubic ft. = 28.32 litres

1 ft/min = 0.018 km/hr

1 pound = 7000 grains = 16 ounces

1 ream = 500 sheets (of paper)

1 Cwt (hundred-weight) = 50 kg = 112 pounds

1 million = 10^6

1 billion = 10^9

1 quintal = 100 kg (100 lbs earlier)

1 league = 4000 m (or) 3 miles (in nautical practice)

1 fathom = 1.8m (or) 6 feet (usually for depth measurements in water)

1 quire = 8 sheets of paper (16 pages)

1 knot = 1 nautical mile per hour

The Pioneers in Engineering and Technology

It is of considerable interest to know something about the pioneers whose names are now frequently used in engineering and technology.

Archimedes	287-212 B.C.	Greek philosopher
H. Pitot	1695-1771	French inventor
D. Bernoulli	1700-1782	Swiss physicist
A. Chézy	1718-1798	French engineer
J.C. Borda	1733-1799	French mathematician
G.B. Venturi	1746-1822	Italian engineer
P.S. Laplace	1749-1827	French mathematician
J.L. Poiseuille	1799-1869	French physicist
W.E. Darcy	1803-1858	French engineer
W.E. Weber	1804-1891	German physicist
W. Froude	1810-1879	British naval architect
J.B. Francis	1815-1892	American engineer
R. Manning	1816-1897	Irish engineer
H. Bazin	1829-1917	French engineer
L.A. Pelton	1829-1908	American engineer
E. Mach	1838-1916	Austrian philosopher
O. Reynolds	1842-1912	British physicist
C.G.P. de Laval	1845-1913	Swedish engineer
L. Prandtl	1875-1953	German engineer
N. Kaplan	1876-1934	Czech engineer
T. von Kármán	1881-1963	Hungarian engineer

The Russian Alphabet and Transliteration

А а	a	К к	k	Х х	kh
Б б	b	Л л	l	Ц ц	ts
В в	v	М м	m	Ч ч	ch
Г г	g	Н н	n	Ш ш	sh
Д д	d	О о	o	Щ щ	shch
Е е	e	П п	p	Ъ ъ	’’
Ё ё	e	Р р	r	Ы	y
Ж ж	zh	С с	s	Ь ь	’
З з	z	Т т	t	Э э	e
И и	i	У у	u	Ю ю	yu
Й й	y	Ф ф	f	Я я	ya

A Little Engineering Humour

1. What did the light bulb say to the generator?
   → I really get a charge out of you!
2. How do you pick out a dead battery from a pile of good ones?
   → It's got no spark!
3. Why do transformers hum?
   → They don't know the words!
4. What did the light bulb say to the electric generator?
   → You spark up my life!
5. What did the baby light bulb say to the mummy light bulb?
   → I love you watts and watts!
6. Why was the free electron so sad?
   → It had nothing to be positive about!
7. What did the Demon say when he ate the nuclear power plant?
   → Shocking!
8. Why did the lights go out?
   → Because they liked each other!

9. Two atoms were walking down the street one day, when one of them exclaimed,

   → Oh, no, I've lost an electron!

   "Are you sure?" the other one asked.

   → Yes, replied the first one, I'm positive!

10. A man with a hearing problem walked into a power plant for a tour. He arrived late and had to join the rest of the group already on the tour. The tour manager was reviewing what he had just told the group. He told the group that they wouldn't move on until they answered this one question: What is the unit of power equal to one joule per second called?

    → The man with the hearing problem hadn't heard the question very well, so he raised his hand and asked What?

11. Engineers calculate all the angles.
12. If we knew what we were doing, it wouldn't be called research.
13. Old chemical engineers never die, they just fail to react.
14. Research is the act of going up alleys to see if they are blind.
15. To err is human, to forgive is divine, but to check—that's engineering.

(*Courtesy*: Journal, *Machine Design*, November 8, 2007, pp. 158-59, Buffalo, New York, U.S.A.)